# Three Stirling Engines You Can Build Without a Machine Shop

An Illustrated Guide

Jim R. Larsen

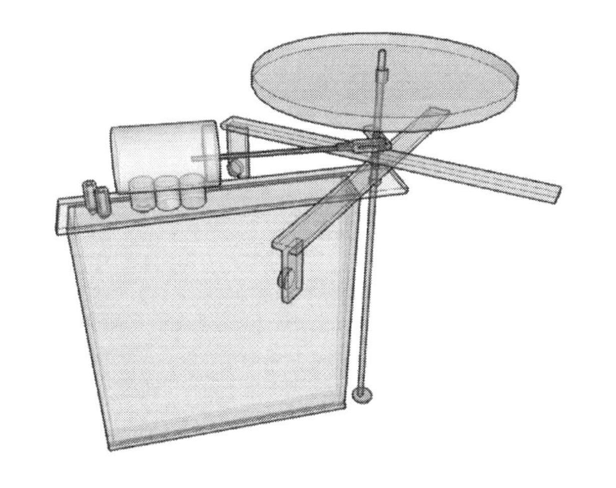

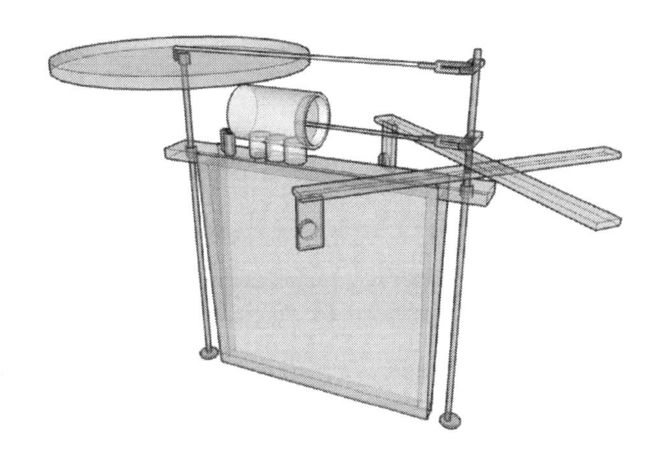

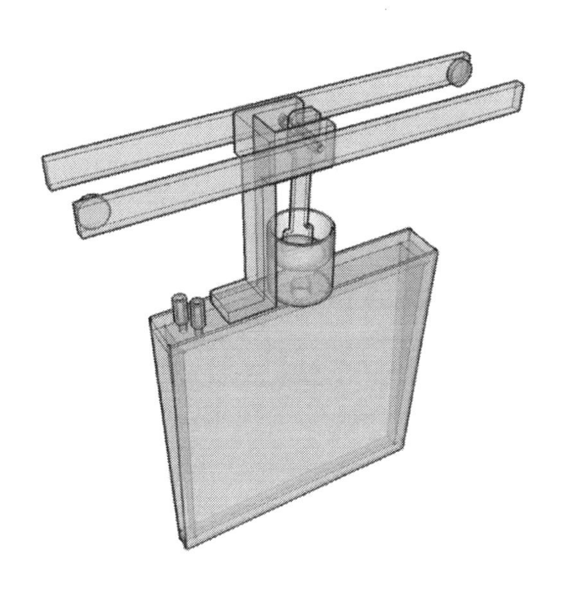

Published by: Jim R. Larsen, P.O. Box 813, Olympia, WA 98507
ISBN number: 1452806578

EAN-13 number: 9781452806570

## About the Author

Jim R. Larsen has a Bachelor of Arts degree in Theology and Pastoral Care from San Jose Bible College (now known as William Jessup University) and is ordained by the nondenominational Christian Church. Jim has a Master of Arts degree in Counseling Psychology from Adams State College in Alamosa, Colorado.

Jim has worked as a Youth Minister, Minister, Counselor, and as a Technical Trainer and Writer.

### *Contacting Jim R. Larsen*

**_YouTube_**
Jim periodically posts videos of Stirling engines on YouTube. His YouTube username is 16Strings. The web address for Jim's YouTube channel is http://www.youtube.com/16strings.

**_Web Page_**
Visit http://Stirlingbuilder.com.

**_Email_**
You can send email to Jim at Jim@Stirlingbuilder.com.

**_Jim's Blog_**
Visit http://woodenmusic.blogspot.com/

All temperatures presented are in degrees Fahrenheit, unless otherwise noted. All dimensional measurements are in inches. Metric conversions are provided in the appendix.

Acrylic sheet handling techniques are recreated with material provided by CYRO Industries, used with permission.

Cool facts about thermal conductivity are quoted from EngineeringToolbox.com. Used with permission.

# Three LTD Stirling Engines You Can Build Without a Machine Shop

## Table of Contents

***Acknowledgements***

This book would not have been possible without the help and support of many of my friends and family. I wish to give a special thanks to my wife Dee Ann for her many hours of help with editing and correcting my manuscript, and for her patience with me as I took over the dining room table to make these strange looking projects. I want to thank my father, Jim Larsen Sr., for teaching me how to be a critical thinker, how to fix things, and how to never be afraid to take something apart to find out how it works. And I want to thank Brent Van Arsdell. Though we have never met, it was your challenge that started this whole ball rolling.

Special thanks to my manuscript review team: Dave Guthmann, Dave Marshall, Mark Chard, Mark Buening, Lori Kesl, Heather Coughlin-Washburn, and Walker Armstrong.

## *Preface*

This book will guide you through the process of building three of your own low temperature differential (LTD) Stirling engines, each built from about $25 worth of materials available in most hardware stores or home centers. When finished, two of them will run from the heat of a warm hand. All three will run from the warmth of direct sunlight, or when set close to a warm light bulb.

The tools needed to build these engines are probably already in your garage. All of the models built in this book were assembled without the use of a machine shop or expensive machine tools.

Many people are on a quest for *Green Power*. They are looking for a way to generate electricity or mechanical motion while leaving little or no carbon footprint. Stirling engines have the ability to run on energy that is currently being wasted by other appliances or by collecting solar energy and turning it into physical work. They don't produce a lot of horsepower, but their advantage is they can be made to run with a very small heat source. The heat coming off of my computer right now is enough to power a small Stirling engine. Finding waste-heat sources and using them to power small scale Stirling engines is a fun and fascinating approach to studying the science of Green Energy.

The engines described in this book are very small in size. They are fascinating to look at and wonderful to watch. Engines this small will not have many practical applications, but they will provide proof of how the Stirling cycle works and will teach you concepts that can be scaled up as far as your dreams will let you go. The "Stirling cycle" is the ability to generate motion by heating and cooling the same captive air (or other gas) in a repetitive, or cyclical, fashion.

I am not an engineer. It would be a stretch of the imagination to call me an amateur engineer. If you start talking to me about drag coefficients, pull-off friction, and differential drag friction equations I won't have any idea what you are talking about. I don't even know if I just used those terms correctly! But I like to tinker, and I like to think I have a basic understanding of how the world works around me. When I was in high school I took every form of wood shop, metal shop, and electric shop class I could get into. My father is a retired technical electrician and he always shares his insights and skills with me. In fact he has prowled through his own garage to find little pieces and parts to be incorporated in my engine designs. So I grew up with tinkering and lots of role modeling for practical know-how. But I am not an engineer!

If you are looking for a technical explanation of how Stirling engines work, or a detailed description of the science behind each component, you will want to include another text in your studies. I will eagerly read those works and absorb all the tricks I can. But when it comes to research, I am a bit more basic. My research usually involves lots of observation, then a flash of inspiration in the middle of the night. The next day I modify my design and try it. If it runs faster (or starts to run at all) I keep the idea. If it runs slower or stops, I try to figure out why and learn from it.

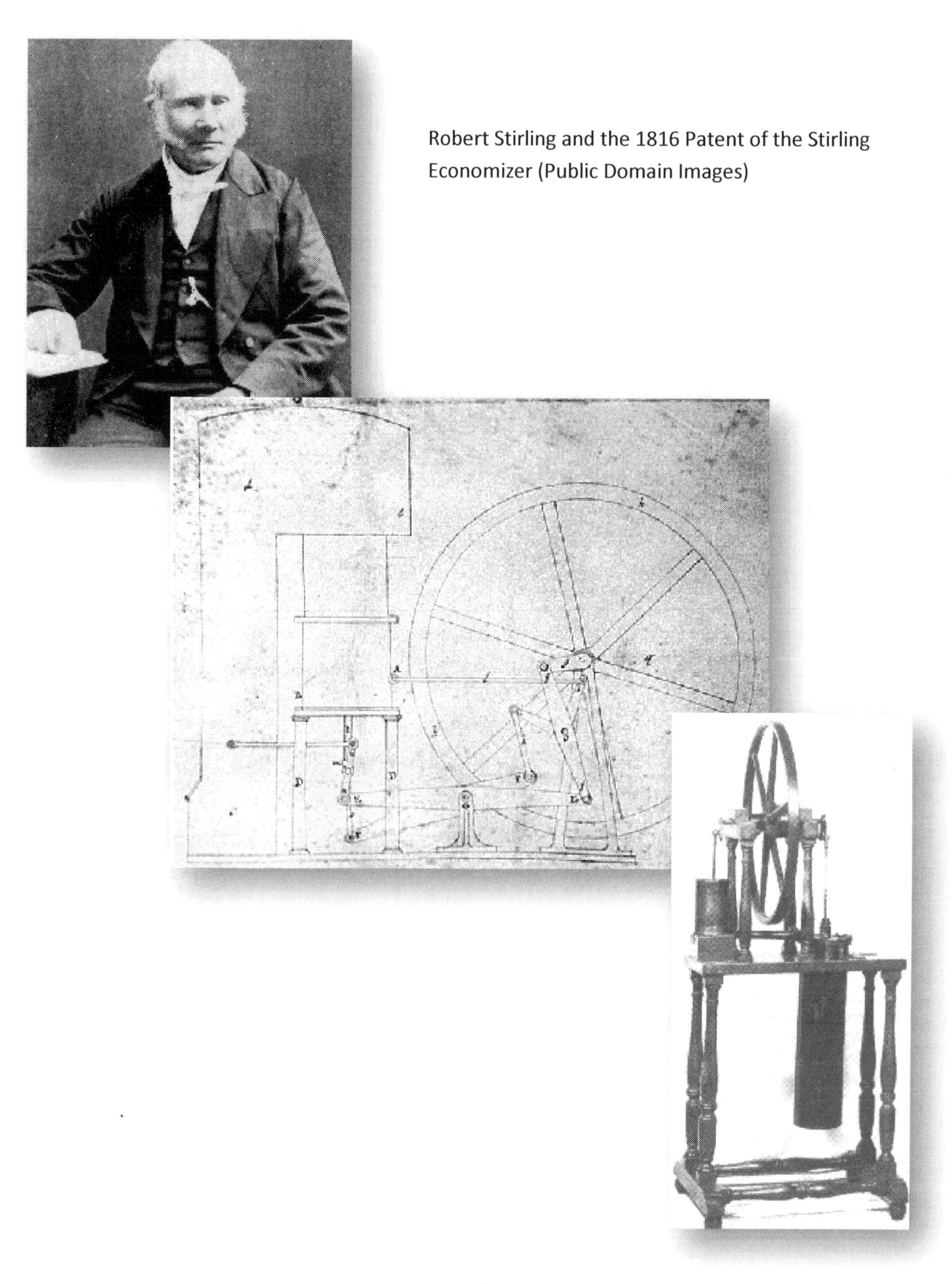

Robert Stirling and the 1816 Patent of the Stirling Economizer (Public Domain Images)

# Chapter 1: My History with Stirling Engines

I was looking for a project. I had recently undergone back surgery and I was anticipating a long recovery with plenty of spare time on my hands. I happened across a catalog with a couple of Stirling engine kits in them and started pondering the idea of buying one of the kits and assembling it. I soon found myself researching and shopping on the Internet.

The first kits I discovered looked like little steam engines. They used a little alcohol lamp for heat and looked like they would spin pretty fast. But in my web surfing I soon discovered an even more fascinating type of Stirling engine. These Low Temperature Differential (LTD) engines would run from small heat sources, such as the waste heat generated by a computer monitor, a television, or even from the heat of your hand. I quickly decided these would be a much cooler toy and started shopping around for plans or a kit.

What I discovered was that most of the LTD engine kits were precision engineered, machined, and relatively expensive. Did I really want to spend $200 to $300 for a toy? I was actually close to taking the plunge and buying one when a friend pointed out that you can build a Stirling engine for almost free by using some pop cans, a coat hanger, and plans you can get off the Internet.

I looked into it and didn't want to take that route at first. These pop can engines were not the impressive LTD designs that could run from the heat of your hand. So as a compromise I decided to build the "free" pop can engine first, then later I could graduate to one of the fancy kits.

## My First Pop Can Engine

That first pop can engine took me about two weeks to assemble. Because of my recovery issues from the back surgery I could only work for 15 to 30 minutes at a time, 2 or 3 times a day. It was a struggle. I was inspired by a video I found on YouTube that showed a similar engine that had been assembled by a young woman named Megan. If Megan could make it work, I could too!

So after I got my first pop can engine up and running, I took a little 30 second video and posted it on YouTube. It was not a huge success by YouTube standards, but it did get me in touch with other people who were on a similar quest to build a fun little engine with little or no investment.

**Figure 1 - My first Stirling Engine was made from pop cans, CDs, and a metal coat hanger.**

My experience building the pop can engine taught me several important lessons. The pop can engine is very forgiving. You won't need ball bearings or a piston. The most expensive thing you will have to buy is the glue.

The pop can design does not require a high degree of precision, although careful attention to detail will ensure your engine works well. I have seen videos of other people's creations that amaze me they actually work. Paying careful attention to detail is important for minimizing friction and increasing balance in the model engine.

The most important thing I learned by building the pop can engine is that the Stirling Cycle really works! I had previously read lots of explanations, watched videos, and talked to experts. But nothing has helped me understand the principles more than building a working model.

## My Next Attempt: The Coffee Cup Engine

I kept thinking about my dream for a low temperature differential engine, but I still didn't want to pay the big price. So I started thinking about how I might make one. Since the pop can engine worked well, and I could see lots of pictures on the Internet of what these things looked like (and videos of how they worked), I thought it would be a fairly easy task.

**Figure 2 - I was hoping this would run from the heat of my hand, but instead it required a large cup of hot water.**

I started drawing plans and creating designs for what I was sure would be a giant breakthrough in Stirling engine history. I was going to make a Stirling engine that would run from the heat of my hand, and I was going to make it with junk I bought at the thrift store. And I would do it all without any fancy machine shop tools. I went to places like the Goodwill store, Salvation Army thrift store, and Value Village. I found bearings in an old pair of skates. I bought old aluminum cookware for the sheet aluminum. And I bought a paint roller handle that would become my drive axle. I used my treasures to fashion a new and original design. I was confident I could run this engine from the heat of my hand.

Well, that is what I hoped for. I found out my engineering skills needed a little improvement. My design worked, but it did not run from the heat of my hand. My first design efforts created an engine that would run when sitting on top of a cup of very hot water. It required a temperature differential of about 100°, which was 10 times more than the 10° temperature differential I was hoping for.

## The Reciprocating Engine

My Internet research eventually led me to the American Stirling Company. While cruising their website for engine kits I ran across an ad that said they wanted a technical writer who would trade some writing of instructional material and plans for engine kits. I shot them a quick email. I am a technical writer in my day job, and I write instructional material for computer software. Now that I was designing my own Stirling engines, I thought I would be a perfect match for them! Well, I didn't end up working for them, but after a few conversations with the owner of the company I was on a mission to complete my quest for a hand built LTD Stirling engine, and I was going to write a book about it at the same time.

Brent Van Arsdell of The American Stirling Company was very helpful. He gave me some very good advice about how to increase the efficiency of my engine designs. He knew a lot about friction. He knew how to calculate its impact, and how to overcome some of the problems it can cause. He didn't really need a technical writer any more, but he did offer me some great encouragement on my efforts to design an efficient hand built engine. He suggested that if I could build a Stirling engine that met these criteria, it would make a great selling book. Here is what my designs had to do:

- The engine had to run from the heat of a warm hand.
- The engine had to be made from common materials that would be easy for anyone to obtain.
- It had to be affordable to build.
- It had to be built without the aid of a machine shop. I could only use the kinds of tools that most folks already have in their garage.

I stumbled across the website of an engineer in France who was working on a similar project. The only problem was the website was entirely in French! I had a brief email exchange with the author and asked him if his website was available in English. He replied, "Sorry, no. It is hard enough to explain in French!"

But I gleaned a lot of helpful hints from his pictures and from his text, thanks in part to tools like the Google Translator. My previous design attempt had been troubled by friction and by pressure leaks. With gems of wisdom from Brent and some neat tricks on the French site I was able to create a design that reduced the number of friction points and eliminated the pressure leaks. My first original Stirling engine was born! The first design was a reciprocating engine. That means the flywheel does not spin continuously in the same direction, but rocks back and forth as it runs. It is efficient enough to run from the heat of my hand.

And remember, I constructed the engine you see here (and several others) while recovering from back surgery. I was under Dr.'s orders to not bend over, don't lift anything more than 10 pounds, and don't pick up anything off of the floor. Those restrictions presented some interesting challenges. Especially when tiny parts would jump off the table and land on the floor.

**Figure 3 – Engine #1: The self adjusting magnetic drive reciprocating Stirling engine.**

I took over half the dining room table for this project, which is where most of my assembly work was done. There were just a few times I felt it was better to move into the shop for a construction step. All of my tools and materials fit in a medium sized storage tub for those times when the table was needed for more mundane thinks, like eating.

You may recall that before I ventured near the LTD engine designs, I started by first building a pop can Stirling engine. Building a pop can Stirling engine is a great way to start, and I highly recommend it. The most popular set of plans on the Internet are posted by the Stephen F. Austin State University in East Texas. You may be able to find the plans if you search the web for "SFA Stirling Engine Project". The pop can engine is easy to build and is a very forgiving design. It will teach you how the Stirling cycle works while giving you lots of success.

My early efforts with the pop can engine and my first attempt at an LTD design taught me several important lessons that went into the design of the three engines here:

- **Friction** is the enemy. The more moving parts you have, the more friction you have.
- **Pressure leaks** have a high impact on LTD designs. The point where the displacer shaft enters the pressure chamber in a traditional Stirling engine creates friction and a leak. Both contribute to a loss of power.
- **Precision** is required, especially at the sliding points like pistons and the displacer pushrod gland.
- **Balance** is critical. An out of balance flywheel consumes more energy than it provides, and balancing them is not as easy as one first thinks.
- Best of all, I learned the **Stirling cycle**.

I learned friction is the enemy. Every point where 2 moving parts touch is a friction point. Even though my friction appeared to be non-existent, it wasn't. The piston arrangement on the Coffee Cup Engine worked, but it had some friction. And the displacer pushrod made noise as it passed in and out of the pressure chamber gland. I knew noise comes from vibration, and vibration comes from friction.

Some of the first flywheels I made would spin freely on their bearings for 4 or 5 seconds and appeared to be working well. I didn't realize how much friction was present until I created my first flywheel on a vertical axle and gave it a spin. I could get it to coast for about 2 minutes! In the micro-horsepower world of model Stirling engines it takes only a small amount of friction to have a huge impact on performance.

I learned pressure leaks will rob you of precious power. A tiny pressure leak will help a Stirling engine "equalize" to match the changing barometric pressure, or help it bleed off a little excess pressure as the average temperature rises in the engine. But with that advantage comes a loss of power. My little engine needed all the power it could muster, especially for the lower temperature differentials I wanted it to run on. Some miniscule pressure leaks seemed impossible to avoid, such as where a moving shaft must pass through the wall of the pressure chamber. I eventually found a way to use magnets to replace that pushrod so now my pressure chamber is completely sealed.

I learned that balance is a difficult science. A typical design has a rotating flywheel that lifts the displacer up and down. You add a little weight to one side of the flywheel in order to compensate for the weight of the displacer. You place the weight opposite of the load so that when the displacer is moving up, the counterbalance is moving down. This makes the displacer motion almost effortless. I have used this technique to successfully reduce vibration and wobble, but I have never been able to completely eliminate it. I needed to create a design that did not require the flywheel to lift anything. If it wasn't lifting anything, balance would not be a problem.

So now the challenge is to apply these lessons to the building of an LTD engine. Lower temperature differential means this engine will have less power than the pop can engine. This means it will have to have more precision, less friction, and a straightforward application of the Stirling cycle. And I have to do this without the aid of a machine shop!

## Applying Lessons Learned

So here is how I applied all those important lessons into the design of my unique Stirling engines:

What if I could use magnets to move the displacer? If I could find a way to do that I would solve two important problems. It would eliminate the friction and the pressure leak caused by the displacer pushrod moving through the top of the pressure chamber.

What if I could eliminate some of the friction points? If I turned the axis of the drive axle from horizontal to vertical I could make it spin like a top. All of the weight could ride on one very small hard point on a hard surface. A single ball bearing can introduce over a dozen friction points. Spinning the flywheel like a top only requires two. The really nice small bearings sell for $17 each. My new design quite literally spins on a penny!

I also realized that with the flywheel lying flat and spinning on a vertical axis, I can turn the pressure chamber on its side too. If I make the displacer so it gently rocks back and forth, the flywheel will never have to lift it, and I will never have to play around with counter balance weights on the flywheel.

I spent quite a few days on the drawing board before I put all the ideas together in a way that I thought might work. I tested the various components as I assembled the engine and made some minor modifications as I went. I became a bit nervous as the engine came close to final assembly. I was very excited about my design until it got close to the time when I would know if it was going to succeed or fail. At that point I started to get a little nervous.

The first time I ran the reciprocating engine I used a 60 watt bulb to heat the warm side, and an ice pack on the cold side. It took very little adjustment, and it worked almost immediately! The next few days were filled with experiments to find the best pushrod position, the best flywheel weight, and the best magnet positions. The adjustability of this little engine makes it great for trying new things. Within a week of that first 'final' assembly, I had it tuned to the point it would run from the heat of my hand. It would also run on ice, or with direct sunshine. This engine was easy to build, relatively cheap, and it worked really well!

At this point I thought I was ready to publish my first book on Stirling engines. I called Brent again, and I soon realized we had a new design requirement. He looked over the design and watched the videos. His expert advice: Build an engine with a rotating flywheel! Brent expressed that for many people reciprocating engines look like an engine that is trying to start but never quite makes it. If I wanted a book that would grab people's attention, I needed to have an engine that behaved in a familiar manner. That is why there are three engines described in the book. The second and third designs are modifications of the first design, each with the addition of a revolving flywheel. Each time I built one of these models I would learn and improve on previous attempts. I made enough improvements in the subsequent models that I eventually went back and rebuilt the first reciprocating engine and included the improvements. So if you follow my instructions here, you will be building the latest and most improved versions.

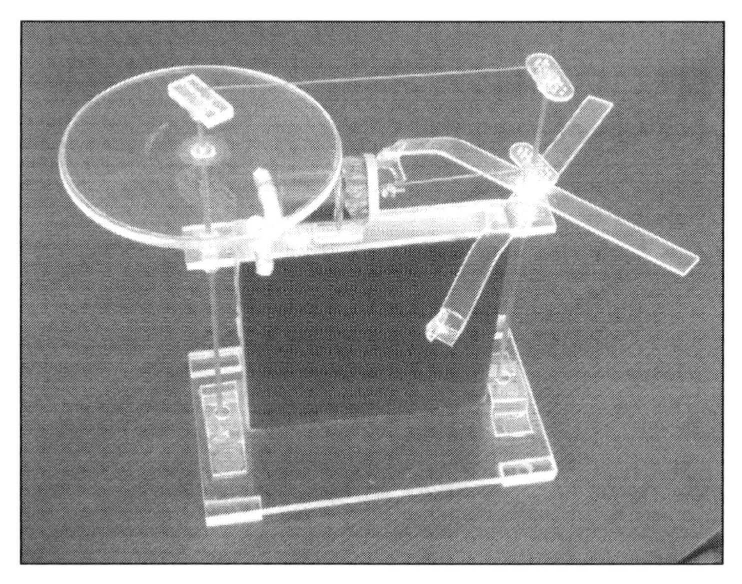

**Figure 4 - Engine #2 has a flywheel that rotates on its own axle.**

Modifying the design for the addition of a revolving flywheel was not difficult to do. Engine #2 takes the flywheel off of the reciprocating drive axle and moves it onto its own rotating vertical axle. The axles maintain the vertical orientation used in the first engine. The flywheel mass still stores kinetic energy, but as a rotating wheel rather than a reciprocator. It is a very efficient design and is fun to operate. It is one of my personal favorites.

Engine #3 was simply an attempt to build a Stirling engine with a rotating flywheel and with as few moving parts as possible. The horizontal axis of the flywheel made it necessary to use bearings. The bearings were made from glass beads instead of steel ball bearings. The beads cost $0.10 each instead of $17.00! The polished metal shaft passing through the bead has a thin coat of silicone lubricant to help fight the friction in the homemade bearings. Engine #3 is not as efficient as the first two because the horizontal drive axle bearings introduce additional friction. I have succeeded in getting Engine #3 to run from the heat of my hand, but only when using helium in the pressure chamber. It will run well on direct sunlight and other low temperature differentials. The glass bead bearings work well enough that you will see them in all three designs.

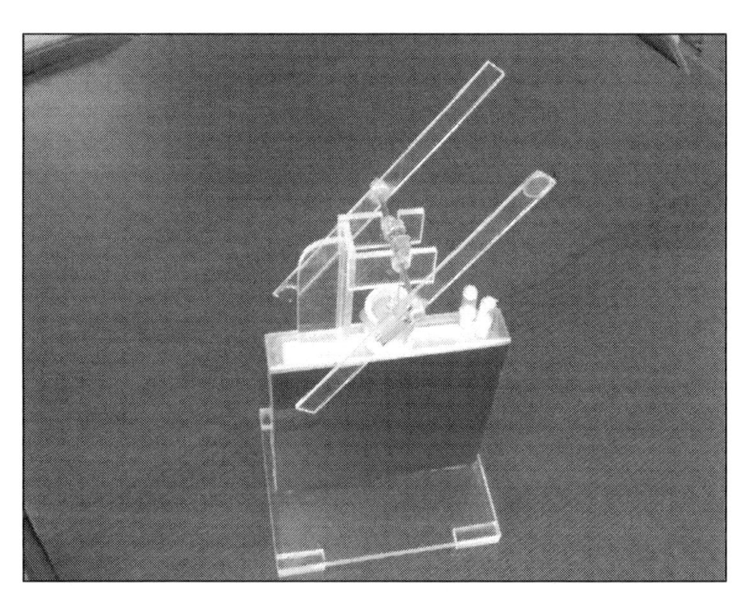

**Figure 5 - Engine #3 is built with as few moving parts as possible.**

Every step in this Stirling journey has been a wonderful learning experience. Building these engines has taught me new tricks and skills for working with metal and acrylic. I have learned how to cut, polish, and paint aluminum, how to cut, bend, weld, and polish acrylic glass, and best of all, how to make several affordable Stirling engines that work from the heat of the hand!

Please note that "heat of the hand" is the goal, but there are environmental factors that will cause results to vary. These engines work by exercising a *temperature differential*. If your hand is not much warmer than the room you are in, you will have trouble creating a significant temperature differential. Engines #1 and #2 will work with a temperature differential down to only 10°. So if your hand is above 90°, and the room is 70 or lower, these engine designs can run from the heat of your hand. Because of the nature of these projects, your results may vary. I can tell you that I have successfully tuned all three of these designs to work effectively from heat of the hand and also from sunlight. You should be able to find my YouTube videos made during my research for the book.

As I mentioned before, the designs you see here are the result of the lessons learned from a series of previous builds. The unique qualities of these designs are intended to reduce friction, reduce pressure leaks, and improve balance. They may even appear to be artistic. Time will be spent discussing how these lessons were learned and why each of the design modifications is important. If you can learn from some of my mistakes, you won't have to repeat them for yourself!

# Chapter 2: A Brief History of Stirling Engines

## Stirling Engine History

> *"...These imperfections have been in great measure removed by time and especially by the genius of the distinguished Bessemer. If Bessemer iron or steel had been known thirty five or forty years ago there is scarce a doubt that the air engine would have been a great success... It remains for some skilled and ambitious mechanist in a future age to repeat it under favourable circumstances and with complete success..."*
>
> Rev'd Dr. Robert Stirling, 1876
> from "Stirling Engines" by G. Walker

The definition of the term "Stirling Engine" has widened over the years since Robert Stirling applied for his patent in 1816. The 1816 patent for which he is famous was not for the invention of the hot air engine. What Stirling invented was a way to make the hot air engine more efficient. This improved efficiency brought a lot of attention to his design and made it possible for the hot air engine to be used in a variety of applications that previously used a steam engine.

Today the term "Stirling Engine" is used as a label for most any closed-cycle hot air engine, even if it does not utilize the unique elements of the Stirling patent. Stirling's patent included the use of what he called the "Economizer". In contemporary designs this is referred to as a regenerator. A regenerator is a heat exchanger that is placed in the airflow between the hot and cool sides of the engine. The regenerator captures heat from the air as it passes through to the cool side of the engine, and then uses that energy to pre-heat the air as it flows back to the warm side of the engine.

The designs in this book do not include a regenerator, so to a purist they would not be a true Stirling engine. However it is widely accepted that since Robert Stirling made the hot air engine a noticeable part of history, we can use the term to address any design that is building upon his work.

The reason for the missing regenerator in small models is mostly an issue of scale. Low Temperature Differential (LTD) engines with their low power output generally see no improvement with the addition of a regenerator due to the added weight and drag they create.

## Almost Famous

The Stirling engine is a design that was almost famous several times in history. Each time it started to find a place in the market, it was surpassed by another new technology that snatched the glow of the spotlight. It happened with the invention of stronger steel, the invention of the transistor, and it is happening again with the introduction of new microchips that produce electrical current from temperature differentials.

## Competing with Early Steam

Rev. Stirling's quote at the beginning of this chapter makes reference to Bessemer Iron. This was the advent of the compound we now know as steel. The introduction of steel made steam engines safer, and led to the development of the internal combustion engine.

Robert Stirling applied for his patent at the same time he was appointed to be a minister in the Church of Scotland at age 25. Many have speculated that he invented the Stirling engine in part because of his concern for the many people in his parish who were being injured or killed by the widespread use of steam engines. Steam engines at the time were made from iron and other soft metals because Bessemer Iron and steel were not yet available. As a result, steam engines could easily explode with devastating results to life, limb, and property. Stirling's patent offered a safer alternative to steam. Stirling engines did not explode like early steam engines did. Early Stirling engines still experienced failures as a result of heat and poor materials, but there are no reports of exploding Stirling engines causing the level of injury or loss of life like there is with the early days of steam.

When Bessemer Iron and steel were developed the steam engine became safer and capable of handling more pressure. The steam engine became lighter and stronger. The safe, reliable, lower horsepower Stirling engine was delegated to less glorious tasks, such as portable water pumps for mining and farms.

Thousands of Stirling engines were produced over the next century by entrepreneurs and inventors such as John Ericsson (1803-1889). A native of Sweden, John Ericsson was an engineering genius who involved himself in the development or improvement of power transmission. He invented the ship propeller and later designed and built the battleship *Monitor* for the Union Navy in the War of the Rebellion. His 1833 "Caloric Engine" developed about 5 HP and used a Stirling style regenerator. Ericsson also designed a 260 foot Stirling powered ship in 1861. The ship did not produce enough speed with the Stirling engine to be able to make it a success, and it eventually sank in a storm. Later, the ship was raised and re-fitted with a steam engine.

The "Caloric Ship" was promoted in the media for its ability to save energy, operate efficiently, and improve safety. Although Ericsson called it a Caloric Engine, it was an adaptation of the Stirling design. When the Caloric Ship was launched, an article in the St. Louis Intelligencer (January 7, 1853) declared, "The reign of steam must date its decline, if not its downfall, from that day." The article suggested that the Caloric Ship could travel on one-fifth of the fuel required by the coal fired steam boilers of the current fleet due to "the wonderful cheapness of its operation." Carrying less coal would mean more cargo, and more profit for the shipping companies. And it would be able to cross the Atlantic on a single load of fuel.

Sadly, the ship was only able to average 7 knots of speed in the open sea. Stirling engines are very efficient, but not very powerful. When the ship was lost to a storm, it was Ericsson himself who suggested the ship be refitted with a steam engine. After the re-fit, his ship went on to gain its own fame during the Civil War, though not for the benefit of Stirling engines. The ship's four small stacks and side-mounted paddle wheels can be distinctly seen in the picturesque group of vessels assembled at the capture of Port Royal, South Carolina, by Commodore S. F. Du Pont.

### Powering the Electric Radio

The second moment in history that brought attention to the Stirling engine design was the invention of the electric radio. Early radios used vacuum tubes. Vacuum tubes consume a lot of power. While it is possible to run a tube radio from batteries, the batteries would not last long. Batteries of the day were bulky, expensive, and not very efficient. Radio was a new powerful medium for communicating over great distances, but many of those whom the market wanted to reach did not have electricity.

The Phillips Company (a large Dutch electrical and electronics manufacturer) began research in the 1930's into Stirling engine design as a way to expand the market for their electric radio sets. The idea was to use a Stirling engine to make a small portable generator that could be used to power the radio. It needed to be economical to purchase and operate.

Their research continued throughout World War II and into the late 1940's. At that time the Phillips Company had developed a marketable version of a Stirling driven generator called the "Type 10" that could produce 200 watts. It was handed off to the Philips' subsidiary Johan de Witt in Dordrecht for production. The first attempts to market the new generator did not happen until 1951. Production costs made it expensive enough that it did not gain wide use as they had hoped. The introduction of the first Transistor by Bell Laboratories in 1947 made possible the invention of the Transistor Radio a few short years later. These new radios used much less energy and would work for long periods of time with just a small battery. Once again the Stirling engine was almost famous.

The Phillips Company continued their research and eventually made several attempts to market their Stirling engine designs, including a Stirling powered vehicle. The only real success they found was using a "reversed Stirling engine" as a source of refrigeration. By using an external motor to rotate a Stirling engine it is possible to cause one side of the engine to get warm, and the other side to get cold.

In recent years the Stirling Design has come back into the spotlight as people have developed in interest in finding energy sources that are ecologically friendly. A lot of attention has been given to finding sources of waste heat that can be harnessed by an LTD Stirling engine and converted to electricity. If you think about it there are heat sources all around us that could be used to create energy. The heat coming off of the laptop computer I am using to type this text is enough to run a small Stirling engine. Computers, monitors, VCRs, and thousands of other electronic devices in use every day are generating heat that could be harnessed and turned back into electricity.

On a larger scale, Stirling engines are being used in experimental solar arrays for generating grid-ready electricity that is being used to provide power to some of the communities where you and I live. New Mexico has a program that has been underway since 2004.

There are scores of small inventors and builders across the globe experimenting with small scale LTD Stirling engines as a way to harvest waste energy and turn it into electricity. A quick search of YouTube for "Stirling Generator" will show a large collection of videos that include Stirling engines that are generating enough current to light a bank of L.E.D.'s or charge a cell phone.

### Harvesting Waste Heat to Make Electricity

We may be able to see history repeat itself one more time with the mighty Stirling engine. Science has recently provided us with the solid state equivalent of the Stirling engine, called a "thermal chip" or "Peltier cooler". Thermal chips work just like a Stirling engine, only with no moving parts. When a thermal chip is placed between two surfaces with a temperature differential, it produces a small electrical current. And, like its Stirling cousin, it can be run in reverse. If an electrical current is applied to a thermal chip, one side of the chip will become cool, and the other side will become warm. These small wonders can refrigerate with no moving parts.

Thermal chips are inexpensive to make using standard semiconductor manufacturing methods. Their efficiency is a matter of some debate. Some call them "highly efficient", while others disagree. It all depends on the standard used for comparison. One application of the Peltier cooler that seems to be catching on is in the manufacture of 12 volt portable refrigerators. If you have ever read the description of one of these car-coolers you may have noticed some familiar talk about temperature differentials. They usually say they will keep the inside of the cooler $30^0$ cooler than the temperature of the outside air.

The practical applications for the Stirling engine may always be usurped by new emerging technologies. But that does not decrease the fascination and fun to be found in creating your own working model engine. There is no replacement for the looks of wonder and amazement people have the first time they see a working Stirling engine. There are many authors who speculate the Stirling engine will always have a place in an oil-starved society that is hungry for energy.

A quick search of Amazon.com reveals over 370 products for sale related to Stirling engines. There is a wide variety of kits, engines, books, and designs for sale. In a similar manner you can search Google and find 144,000 entries for "Caloric Ship" and about 593,000 entries for "Stirling Engine". The Stirling engine is not about to fade into history just yet. It is still showing promise as an alternative method for turning heat into work. Those Google search results are about double of what they were just 3 years ago.

# Chapter 3: The Stirling Cycle Explained

There are explanations of how Stirling engines work all over the web, and in many good science and engineering textbooks. Reading these texts and studying the diagrams will help with the basic concepts, but nothing can replace the learning experience that can be gained by building your own working model. These projects will be easiest for those who have a basic knowledge of how a Stirling engine works.

## A Basic Explanation of the Principle

The fundamental principal at work in a Stirling engine is the fact that air expands in volume when it is warmed, and decreases in volume when it is cooled. The engine has a sealed chamber that is holding a measured amount of captive air. There is a device inside the engine (called a displacer) that moves the air back and forth from the warm side to the cool side of this captive chamber.

- The air warms and expands when it is on the warm side of the engine.
- This results in an increase in pressure inside the chamber. The increased pressure presses on the drive piston.
- The drive piston rotates the drive mechanism and causes the displacer to move the air to the cool side of the engine.
- The air contracts when on the cool side, and the drop in pressure pulls on the drive piston.
- The drive piston rotates the drive mechanism and causes the displacer to move the air to the warm side of the engine and the cycle repeats itself.

This creates two power strokes per cycle, which can make for a smooth running engine.

The piston is usually connected to a flywheel. The flywheel acts as a kinetic energy storage device. It captures a little extra energy from each stroke of the piston and uses that to carry the engine through those brief moments when there is no pushing or pulling going on.

The difference in temperature between the warm side and the cool side of the engine is called the "temperature differential". The engines in this book are Low Temperature Differential (LTD) designs. I am defining "low temperature differential" to be a difference in temperature of 20° or less. Some of the best designs on the market currently advertize they will run down to a 7 degree temperature differential. That is impressive! The engines described here will work with temperature differentials down to 10° (measured on the surface of the engine). That is approaching the efficiency of engines produced in a machine shop.

Air is not the only gas that expands and contracts as it changes temperature. The lighter a gas is, the more it is affected by changes in temperature. Helium and Hydrogen both outperform air when used in a Stirling engine. The engines in this book are designed to run with either air or with helium. Helium is easy to come by and can often be had for free. Using helium in the pressure chamber will cause a noticeable increase in engine performance.

## What Makes These Designs Unique?

Some who view the first engine in action might react to the reciprocating flywheel and comment with something like, "That's not an engine. It doesn't turn." When you think about it, the heart of most engine designs is a piston reciprocating up and down in a cylinder. That action is converted into a rotation motion through the use of a crankshaft. Unless your engine is a rotary or a turbine, it is probably a reciprocating engine of some sort. Engine design #2 is a modification that adds a spinning flywheel for those who want a more traditional output. Adding the flywheel adds a couple of moving parts and a couple of friction points, but the added friction is more than compensated for by the presence of the kinetic energy stored in the spinning flywheel.

These designs have been simplified through the addition of a magnetic drive to move the displacer back and forth horizontally. Traditional Stirling engines use a shaft passing through a gland (a pressure seal) to move the displacer vertically. The tolerance on this joint is so tight in some designs that lubrication will actually make it run slower. A good displacer gland is tight enough that air does not leak out, yet loose enough that it does not cause friction on the pushrod. Tolerances that tight are very difficult to reproduce with hand tools in the ordinary shop. The magnets eliminate the need for the gland and shaft, thereby eliminating a source of friction and a potential pressure leak. Turning the pressure chamber on its edge means the displacer can rock back and forth and does not have to be lifted by the engine.

Another feature of these designs is the ease with which everything is adjustable. I attempted to build adjustability into every part that might possibly need it. This was originally done because the engine was a "proof of concept" and it was desirable to have the ability to make tweaks and changes without having to manufacture new parts. It has proven to be a very useful feature. It is necessary to make small adjustments when the barometric pressure rises or falls, or when operating with various heat sources and pressure differentials. For instance, if you decide to run the engine on ice, it will drop the average temperature inside the engine. This causes the pressure to drop, and changes the relative position of the power stroke of the piston. The engine won't run with the same settings used for warmer temperatures unless you open a vent to equalize the pressure chamber, or in these designs, just reach up and give the crankshaft connector arm a little twist.

### *A Description of Engine #1*

The first engine design in this book takes a modified approach to the Stirling cycle. This is a reciprocating engine with no physical coupling between the drive piston and the displacer. Here is a description of how this engine works:

There is an air-tight pressure chamber like other Stirling engines. And like the others, it has a displacer that moves the air from the warm side to the cool side and back again. When the air heats and expands it presses on the drive piston which rotates the flywheel in a counter-clockwise direction.

The flywheel mechanism includes two small powerful magnets, and there is another magnet on the displacer. The magnets on the displacer are set to oppose the magnets on the drive mechanism, so as the flywheel magnets approach the displacer, the displacer is pushed away.

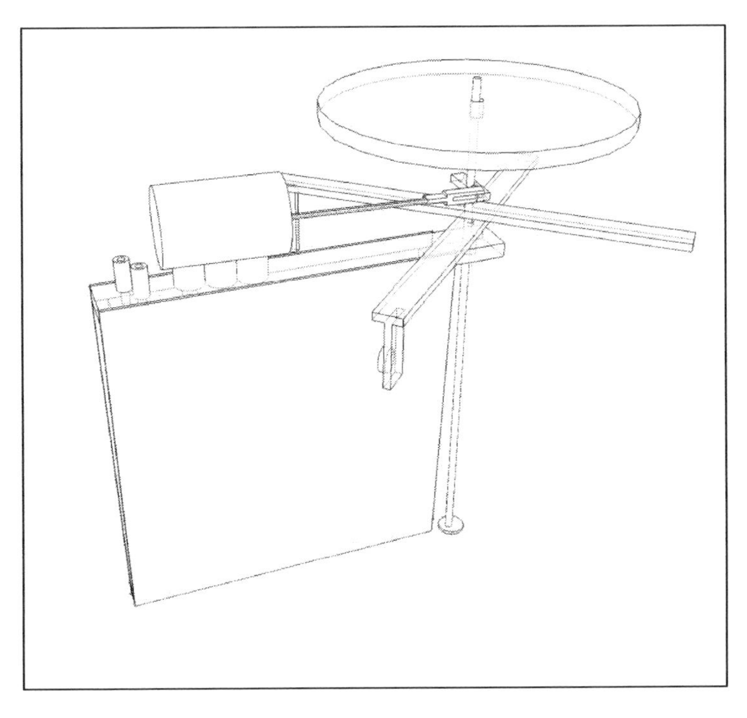

Figure 6 – Engine #1 uses magnets to move a vertically oriented displacer.

When the air is on the warm side of the engine the flywheel will rotate counter-clockwise until the magnet pushes the displacer which moves the air to the cool side of the engine. The momentum of the flywheel causes the two opposing magnets to press a little closer before changing direction and springing back in a clockwise direction.

The air inside the pressure chamber begins to cool and pull on the drive piston as soon as the displacer moved the air to the cool side. The flywheel is now moving clockwise as a result of the piston's effort and the stored energy it picked up from the magnetic spring. It will continue to rotate until the opposing force of the second magnet moves the displacer, compresses the other magnetic spring, reversing the flywheel direction and continuing the cycle.

The magnetic link between the flywheel and the displacer gives this engine some unique self-adjustment features. In a classic design there would be a flywheel with a crankshaft. The crankshaft restricts the motion of the drive piston so every stroke is the same length, no matter how much pressure is applied. These engines will run faster when the temperature differential increases, but there is a lot of unused pressure at higher RPMs. This is because even though there is more air pressure inside, the displacement of the drive piston never changes.

With the magnetic drive and a reciprocating flywheel, things function a little differently. As the temperature differential and air pressure increases in this design, the stroke of the drive piston is not only faster, it becomes longer! As the internal engine pressure increases, the displacement of the drive piston increases. It is thereby able to put more of the internal pressure to work and convert it to physical energy. It is like a car with a variable sized engine. When power supplies are low, it becomes a small displacement engine. The magnetic drive adjusts the piston stroke to be short and you are able to purr along on your economical small displacement engine. If you need more power, you step on the gas (increase the temperature differential) and the engine responds by picking up speed and taking long powerful strokes. The displacement of the drive piston doubles to compensate for the extra power. You now have a powerful large displacement engine! This engine can function under a variety of temperature differentials with very little adjustment. You can also manually increase the displacement of each power stroke by pulling the magnets farther apart. It is a very quick and easy adjustment to make.

Another key difference that adds to the efficiency of this design is the motion of the displacer. A classic design displacer moves up and down on a pushrod connected to a crankshaft. It is always in motion except for the brief moment when it is changing directions. In this new design, the displacer moves quickly from one side to the other, then pauses in the maximum position until the last possible moment when it is quickly moved back to the other side. My testing indicates the most power is being generated when the displacer is at the far end of its motion path. The binary action of this engine keeps the displacer in the full-power position much longer than the classic crankshaft design. The result is that the air heats and cools faster, creating more power.

The binary action of the displacer makes it possible to move the timing of the drive piston and the displacer much closer together. In fact, it is almost instantaneous. In a traditional Stirling engine design, there is a 90 degree offset on the crankshaft between the displacer and the drive piston. This compensates for the slower, constant motion of a displacer that is linked to the flywheel by a crankshaft. The drive piston in these designs does not have to be 90° behind the displacer because the displacer moves almost instantly. The periods between power strokes are shorter because the power strokes are longer. The drive piston is listening for pressure changes and can be seen to change directions with the motion of the displacer. Synchronization is automatic. The only thing we have to adjust is the length of throw for the piston and center point of the power stroke.

If you have never built a conventional Stirling engine you probably don't have a clue about what I just said! So if that is you, pour yourself a nice cup of coffee, sit down in a comfortable chair, and read on.

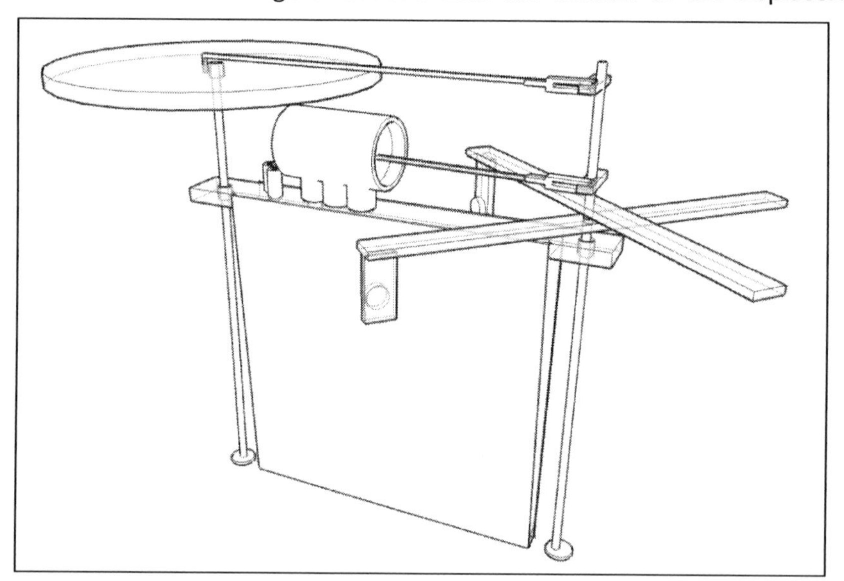

Figure 7 - Engine #2 adds a rotating flywheel on a second vertical axle.

### *A Description of Engine #2*

The second engine design presented here is very similar to the first. The only difference is the flywheel. It still has a magnetic drive that pushes the displacer back and forth to move air from the warm side of the engine to the cool side. The reciprocating magnets are connected to a flywheel that spins rather than reciprocates. This engine has a fixed stroke length and is therefore not self-adjusting like the first model. The two designs are very similar. When you see the plans, you will probably notice that engine #2 is actually just engine #1 with a few additional parts added.

### *A Description of Engine #3*

The third engine provides a pair of rotating arms that function as the flywheel and as the magnetic driver for the displacer. This design minimizes friction by using only three moving parts: the flywheel, the push rod shaft, and the displacer. All three engines utilize a vertically aligned pressure chamber with a rocking displacer. It takes very small amounts of energy to move the displacer back and forth within the pressure chamber when configured in this manner. The vertically aligned displacer does not place any weight on the crankshaft, so the flywheel does not need to be balanced to compensate for the weight of the displacer. There is only a small counterbalance needed to offset the weight of the crankshaft.

This is the most challenging engine of this collection. It must be built with extreme attention to detail, and you must be willing to tinker with adjustments. When setup properly this engine will sit in the sun and spin for long periods of time. I have successfully run this from the heat of my hand, but only when the pressure chamber was filled with helium.

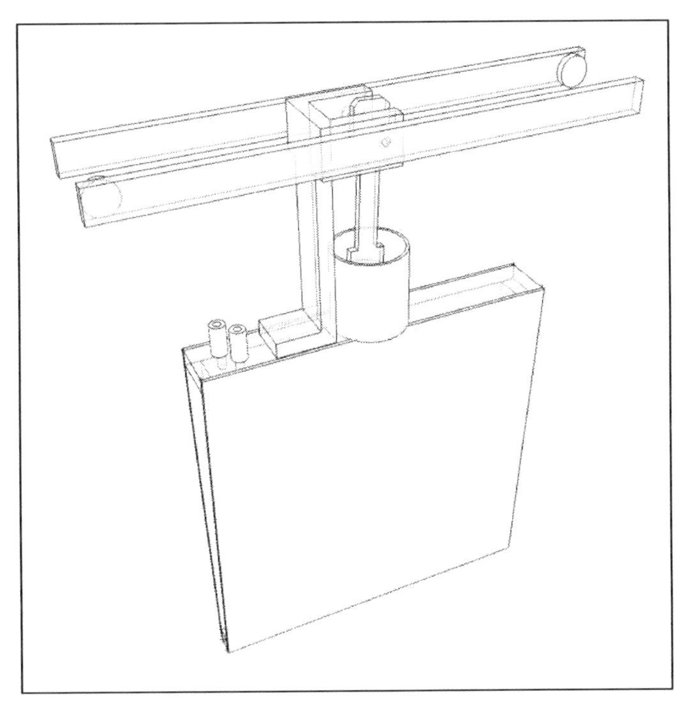

**Figure 8 - Engine #3 has only 3 moving parts.**

# Chapter 4: What Makes a Good Stirling Engine?

## Materials

A good Stirling engine contains a combination of materials, each selected for performance and efficiency. Some of them are excellent thermal conductors, while others are good thermal insulators (in other words, poor thermal conductors). Heat conduction is the flow of internal energy from a region of higher temperature to one of lower temperature by the interaction of the adjacent particles (atoms, molecules, ions, electrons, etc.) in the intervening space.

### *Conductors*

The Stirling engine has a sealed air chamber that has both a warm side and a cool side, with a heat source that is on the outside of the chamber. There are two points in a Stirling engine where heat conduction must be efficient in order for the engine to work well. These two points are the large surfaces of the pressure chamber. The warm side of the pressure chamber needs to conduct heat from the external source (a warm hand, or sunlight) to the air inside the chamber on the warm side. The metal surface that makes up the cool side of the pressure chamber must be able to efficiently conduct heat from the internal captive air of the motor to the external environment in which the motor is running.

If the warm side cannot collect and conduct heat efficiently, the air inside will not expand to run the engine. If the cool side cannot conduct heat away from the pressure chamber, it will become warm on both sides and the motor will cease to operate. (Keep this in mind for later when you are trouble-shooting problems with a Stirling engine.)

Aluminum is chosen for many Stirling engine projects because it is readily available and affordable, and it is a good conductor of thermal energy. If you have ever compared a plastic soda container to an aluminum container in the same refrigerator, you may notice the aluminum feels much colder than the plastic. That is because the plastic is a poor conductor of heat and is insulating your fingers from the cold beverage inside.

Www.engineeringtoolbox.com offers this list of thermal conductivity for various substances:

**Thermal Conductivity- k - (W/m K)**

| Material/Substance | Temperature 25°C | Material/Substance | Temperature 25°C |
|---|---|---|---|
| Silver | 429.00 | PVC | 0.19 |
| Copper | 401.00 | Wood, oak | 0.17 |
| Gold | 310.00 | Hydrogen | 0.17 |
| Aluminum | 250.00 | Wood across the grain, yellow pine | 0.15 |
| Magnesium | 156.00 | Helium | 0.14 |
| Zinc Zn | 116.00 | Plywood | 0.13 |
| Brass | 109.00 | Softwoods (fir, pine ...) | 0.12 |
| Nickel | 91.00 | Wood across the grain, white pine | 0.12 |
| Platinum | 70.00 | Polypropylene | 0.1 - 0.22 |
| Tin Sn | 67.00 | Wood across the grain, balsa | 0.06 |
| Iron, cast | 55.00 | Snow (temp < 0°C) | 0.05 - 0.25 |
| Carbon Steel | 54.00 | Paper | 0.05 |
| Steel | 46.00 | Balsa | 0.05 |
| Lead Pb | 35.00 | Fiber insulating board | 0.05 |
| Stainless Steel | 16.00 | Corkboard | 0.04 |
| Carbon | 1.70 | Fiberglass | 0.04 |
| Porcelain | 1.50 | Styrofoam | 0.03 |
| Corian (ceramic filled) | 1.06 | Cotton | 0.03 |
| Glass | 1.05 | Plastics, foamed | 0.03 |
| Pyrex glass | 1.01 | Polystyrene expanded | 0.03 |
| Epoxy | 0.35 | Cotton Wool insulation | 0.03 |
| Nylon 6 | 0.25 | Air | 0.02 |
| PTFE | 0:25 | Urethane foam | 0.02 |
| Acrylic | 0.20 | | |

Calculations of conductivity take into account the temperature differential, distance (thickness), and area.

As you can see from the table above, there are few metals that beat aluminum in thermal conductivity. Those that are superior to aluminum are silver, copper, and gold.

### *Insulators*

The Stirling engine will be more efficient and will have a decreased chance of overheating if there is a thermal break between the warm side and the cool side. An insulating thermal barrier prevents the heat on the warm side from being conducted to the cool side through the body of the engine itself. The highest efficiency will happen when it is the captive gas that transfers heat from one side of the sealed pressure chamber to the other. Insulators are those materials which are known to have a poor ability to conduct heat.

The short sidewalls of the pressure chamber need to be made of a material that will withstand the working environment of the engine, while also inhibiting the flow of heat through the engine structure itself. As you can see in the table above, several kinds of wood and plastic have a very low level of thermal conductivity. If they have sufficient structural strength and can be made airtight, they will make good short sidewalls for the pressure chamber. So yes, we could make our pressure chamber from pine wood and aluminum. But the clear acrylic is going to look much nicer, and will expose the internal working parts of the pressure chamber.

For these reasons, the engines you see here have aluminum surfaces with acrylic side walls on the pressure chambers.

## Low Friction

Friction can never be taken for granted. In the micro-horsepower world of the LTD Stirling engine it takes very little friction to bring the entire project to a standstill. Friction happens at the points where moving parts touch (including moving parts that touch the air). There are two principles for reducing friction employed by these designs:

1. Reduce the number of moving parts.
2. Minimize all remaining friction.

Efforts were made in each of these designs to eliminate friction without using ball bearings. The main reason is the cost of the ball bearings will add up quickly and make it too expensive for many home experimenters. One of the conclusions of the French research was that glass beads can work as well as ball bearings in some conditions. There is no significant gain in efficiency of a rotating joint when using ball bearings if the load on the joint is less than 10 grams. The glass beads cost $0.10 each. The bearings can cost $17.00 each, or more. Use some of the money you saved on bearings and buy yourself a good cup of coffee. But save the lid. We are going to need that later to make some parts.

There is another source of friction many fail to take into consideration, and that relates to atmospheric drag. Engine parts moving through the air will encounter resistance from the air. That resistance amounts to friction and will rob you of precious micro-horsepower if you let it.

Another form of friction related to the atmosphere is known as pull-off resistance. You might assume you want the displacer inside the pressure chamber to travel all the way to one side until it rests flush against the metal chamber wall. This will actually reduce the efficiency of your engine, possibly even making it so it cannot run at all. This is because the energy needed to pull the displacer off the flat surface of the wall is greater than the energy gained by moving it all the way over to the wall.

When two flat objects are placed completely flush against each other, there is in effect an atmospheric vacuum between them. Press a flat square of cardboard against the ceiling with a broom handle and then remove the support. You will see that it will stay stuck to the ceiling until the vacuum behind the cardboard is replaced with air from the room. The cardboard doesn't fall immediately because the air pressure in the room is holding it against the ceiling. The air quickly leaks in behind the cardboard and it soon falls to the ground. It falls because the vacuum has drawn air in and equalized the pressure on all sides (and of course, gravity plays a part).

The displacer in your Stirling engine will act in a similar way if it comes into flush contact with the metal wall of the pressure chamber. It will stick for a moment and cause an increase in drag as air rushes in to fill the low pressure area between the displacer and the pressure chamber wall. There are several ways to combat this friction. All of them involve creating a path for air to move around or through the displacer as it moves from side to side in the engine. If it stops about 1/16" away from the flush surface, that space will provide a path for incoming air when the displacer has to move. Air paths can also be created by putting grooves in the surface of the displacer, or by making it out of a porous material that will allow air to flow through it when necessary.

The reason the displacer in these designs has a vertical orientation is because it requires less mechanical effort than lifting it vertically. This reduces the amount of friction present in the overall design of the engine.

A good Stirling engine designer will look for friction everywhere and take action to reduce or eliminate it.

## Precision and Balance

If you watch the YouTube videos of the pop can Stirling engines that are made by teenagers for their science fair, it quickly becomes obvious that a Stirling engine will run even if it is not well balanced. Those little engines have a relatively high temperature differential, which leads to more pressure for the drive piston, and more forgiveness in design and assembly. Low Temperature Differential (LTD) engines do not create as much horsepower and have far less tolerance for error.

Every engine has to be tuned for its optimal operating range. The temperature differential and the volume of gas inside the pressure chamber will determine how much displacement is needed in the drive cylinder. The diameter of the drive cylinder will determine the length the piston will travel, which in turn determines how long the crankshaft offset needs to be. These are just some of the examples of why precision is important. If the piston cannot use all the power the chamber is making, energy is wasted. If the piston has too much displacement for the available power, the engine won't run.

One helpful method for making all these variables come together in a hand built engine is to make everything as adjustable as possible. That is why you will notice that these designs include things like crankshafts that can be adjusted for length and position, adjustable pushrods, and easily adjustable flywheels.

If you want your engine to run from a low temperature differential, such as the heat of your hand, it will have to be tuned precisely for those conditions. Strive for precision and balance, and build in the ability to make adjustments.

## Simple to Build

The whole point of this book is making a Stirling engine simple to build! A good Stirling engine design for this project is one that is easy for the reader to replicate themselves. Any model making that has you manufacturing all of your own parts from scratch is not a beginner's project by definition. The book assumes some hobby or woodworking experience. The detail on working with acrylic and working with aluminum is provided for those who may need some additional information. But throughout the design process I have learned to capitalize on concepts that will make the project easier for you to replicate.

Traditional LTD Stirling engines have a round horizontal pressure chamber. After building round chambers and square ones, I find the square much easier to build. The magnetic drive is also much easier to get up and running than a gland and a pushrod can be.

The materials for these projects are not hard to find. Whenever possible, the less expensive option was chosen. Each of these designs will cost about $25 to build, and will leave you with leftovers to start another engine project.

### *View Videos Online*

If a picture is worth a thousand words, then a video is worth millions. The diagrams and discussion in this text will make much more sense if you can view some video of these engines in motion. There are a couple of ways you can access videos of the engines created in the writing of this book. I have posted a collection of Stirling engine videos on a YouTube page. You can find it by surfing the Internet to http://www.youtube.com/16strings. My user name on YouTube is *16strings*. You can search YouTube for "16strings Stirling Engine" and find several video clips of these and other engines I have built and tested. Feel free to post a video response with footage of your working engines so I can see your work!

I have developed a webpage in support of this book that will give you and others in the Stirling engine community a chance to stay up on the latest designs and modifications. You will find the website online at http://Stirlingbuilder.com.

# Chapter 5: Working with Aluminum

Working with aluminum is relatively easy. The soft nature of this metal makes it easy to cut. It can be sanded almost like wood. It can be polished to a mirror like finish or painted. The biggest challenge for me has been finding a reliable source for the aluminum that I can recommend to my readers.

My last batch of aluminum came from the surplus store of the Boeing Corporation for $1.50 per pound. That store is now closed, but they said they will be offering much of their surplus material for sale online. I have noticed several hardware stores carry a small selection of sheet metals that includes aluminum sheet and polished aluminum tread plate. The tread plate looks like it would make a handsome project, but it is a little expensive, about $30 for a 2' by 4' sheet.

I have made a Stirling engine from aluminum cookware I bought at the local Goodwill store, but then I found it was just as cost effective to go to the model shop and buy a 6" by 12" piece of sheet aluminum for $7. One cut across the middle of that and you have the sides of your pressure chamber!

The aluminum you choose will have an impact on how much finish work you will need to do. If you purchase the 6" by 12" piece from the model shop, it will already be flat and burnished to a nice shine. You will simply need to cut it to size and paint it black. But if you try to salvage some sheet aluminum from an old frying pan like I did for one of my early engines, you will need to know a little more about how to cut, flatten, and burnish the metal. I have written the next section with that in mind. Not everyone will need this level of detail, but it is here for those who do.

## Cutting

As I mentioned earlier, aluminum is soft and easy to cut. The challenge is cutting the aluminum without bending your material to the point it cannot be used. Tin snips cut aluminum well, but the problem with tin snips is that the material must bend up or down on one side of the cut or the snips can't move forward. My use of tin snips usually leaves me with one flat part, and lots of curled metal that gets thrown on the scrap heap.

The pros use a sheer. It is like a giant paper cutter with a long blade that chops the metal to size with a vertical sheering motion. I don't expect you to have one, but access to one may be easier than you think. There are lots of machine shops and small businesses that use a metal sheer. Companies that install heating and air conditioning usually have a sheer in their arsenal of tools for making ductwork. Metal fabrication and welding shops will often have a sheer for working with sheet metal. Industrial supply houses that sell sheet metal products may have a sheer in their shop. The only aluminum parts we need for each of these designs are two pieces of aluminum sheet cut into 6" x 6" squares. You may be able to find someplace that will sell the metal cut to order. You may also be able to contact one of these other shops and have your metal cut into 6" squares for a small price.

If you prefer to do it yourself, there are several ways to cut sheet aluminum that will not cause it to bend or distort. Metal cutting saw blades are available for hand saws and power saws. If you have a jig saw or a band saw, you can make fast work of your aluminum cutting tasks. Don't attempt to cut aluminum with a blade designed for wood. You risk damage to your tools and to your body by doing so. You can buy metal cutting blades for just about any power saw. If you have a saber saw or a jig saw, you might give it a try. I am not brave enough to cut small metal parts on a table saw. If the parts were to bind and get away from you, the results could be catastrophic and harmful to living creatures!

A hack saw will work. There are no curves to cut, but some of the cuts are long enough that the hack saw will be stretched to the limits of effectiveness.

My favorite tool for cutting sheet aluminum is a nibbler. One of these will come in handy if you are planning to salvage your aluminum sheet from old cookware or some other manufactured item. Nibblers are designed to cut straight or curved lines in sheet metal without causing serious distortion on either side of the cut. There are two basic styles of nibbler. One removes small rectangular pieces of metal as it "nibbles" its way through the sheet. The other removes a strip of metal that curls into a coil as you cut. Both types of sheers make a cut that is wide. Some material is lost as the cuts can be 1/8" to 1/4" wide.

## Flattening

If you scrounge for aluminum you may find some sheets that are not perfectly flat. Small imperfections can be corrected by hammering the aluminum of a hard flat surface. Be careful. It the flat surface is rough, like concrete, it will leave a rough surface on the face of the material. If the hammer blows are too heavy, you will create thin spots in the metal that you don't want.

I have successfully flattened aluminum by laying it on my table saw and striking it with the flat surface of a heavy wood block. This seems to work well for removing minor imperfections without creating thin spots or hammer marks in the metal.

Cutting with a nibbler can leave a slight curl to the edge of your aluminum sheet. Once again, place the metal on a hard flat surface and lightly tap the edge with a hammer or a heavy wood block.

## Finishing

### *Color Choices for LTD Engines*

The first time I made one of these engines I painted the aluminum plate for the sides of the pressure chamber. Engines 2 and 3 initially were not painted. Paint improves aesthetics, and will definitely improve the engine's operation when powered by sunlight. The insulation provided by the paint is more than compensated for by the increased ability to absorb heat.

I performed a test to see if the engine that was painted black would outperform one that was not painted. I placed engine #1 (painted black) and engine #2 (unpainted) side by side in the sunshine on a 70 degree day. I pointed an optical thermometer at the sunny side of the engines and found the one painted black was over

100° in a few minutes, while the polished aluminum of engine #2 stayed very close to the 70 degree temperature of the air. This test convinced me all my engine bodies needed to be painted black! If I had conducted this test sooner, I would have painted all the sheet aluminum before it was assembled. You may notice in some of the pictures of engines #2 and #3 that they did not receive their paint until partly assembled.

Black is better at both absorbing and radiating heat. Several people have advised me that both the hot side and the cold side of the engines should be painted black. The black on the cold side will help radiate heat out of the engine. At the time of this writing I have not tested black paint for the cold side of the engine.

Aluminum accepts paint well. Wash all the parts with mild soap and warm water before applying a primer and top coat. I have used spray enamel with great success. Make sure the paint is recommended for metal surfaces by reading the label carefully.

### *Be Careful*

It is always a good idea to wear gloves and eye protection when cutting metal parts.

## Chapter 6: Working with Acrylic

Building some of your engine parts from acrylic sheet will add a huge *Wow!* factor to your project.

When people first see a Stirling engine in motion there is a lot of wonderment about how it works. Is this perpetual motion? It is fun to see the initial reactions, the head scratching, and the amazement as they hear the science behind the motion.

That *Wow!* factor is even greater when people examine your work in motion and try to figure out how you made it. Using clear acrylic to manufacture your own parts is easy to master, and gives your project a strong visual appeal. Since it is not a common building material for most home craft projects, it will make your projects really stand out. I have built Stirling engines from pop cans and glue, and I have built them with sheet aluminum and clear acrylic sheet. The pop can engines go together quickly, but the clear acrylic flywheels and moving parts are quite beautiful in motion.

For years I always referred to acrylic sheet as Plexiglas. Plexiglas, however, is simply a brand name for one company's formulation of acrylic sheet. There are actually a broad variety of manufacturers and formulations. One could get quite particular if they wanted to about which type is exactly the best for our application. Acrylic sheet comes in a variety of thicknesses, from less than 0.10" to an inch or more thick. It can be clear, opaque, or translucent colors. Different formulations have different melting points, which can affect how easily they can be bent or molded into specific shapes.

Acrylic sheet can be cut, shaped, sanded, polished, bent, welded, glued, drilled, and basically manipulated into almost any shape imaginable. Once you start to work with it to make your own custom parts, you will probably start looking for other fun things to make with it. It is just plain fun to work with!

For the projects in this book, we will need to learn just a few of the fundamentals for working with acrylic. This will include **cutting**, **bending**, **drilling**, **welding** (gluing), and **finishing**. The best place to go for information about working with acrylic is to go directly to the manufacturer. Be very careful about advice you find in forums on the Internet. Some of the advice I have read in forums is in direct conflict with the safety recommendations of the manufacturers of acrylic. Several forums have suggested heating acrylic sheet in your kitchen oven for bending and forming. If that bad advice is followed, it could result in the release of toxic fumes, fire, or serious bodily injury. So please be careful and make sure you read, understand, and follow the basic safety guidelines.

One of the largest manufactures of acrylic sheet products is CYRO Industries of Parsippany, NJ, USA. They are the manufacturers of ***ACRYLITE® acrylic sheet*** products. They offer a large library of material safety datasheets. Safety information is accessible online. The information contained here about working with acrylic sheet is from CYRO Industries and is used with their permission. As you read through the next section of the book, please keep in mind that these are not the instructions on how to make a Stirling engine. We will get to that shortly. This is a primmer in how to work with Acrylic.

### *General Handling Instructions and Guidelines*

### *Characteristics of Acrylic Sheet*

Acrylic sheet products behave differently from other types of material. Learn what it can do. . . and what it can't. Learn how to care for it. You'll be sure to get the best results. Some of the characteristics mentioned here have no impact on Stirling engine designs, but are good things to know about acrylic sheet products.

**Expansion and contraction:** Like most plastics, acrylic sheet responds to temperature changes by expanding or contracting at a far greater rate than glass. When acrylic sheet is used for outdoor window glazing, the sheet is cut approximately 1/16" per running foot (or 0.5 cm per running meter) shorter than the frame size to allow for expansion. For most low temperature Stirling engine designs this is not a problem because the parts are small and the temperatures relatively low.

**Flexibility:** Acrylic sheet is much more flexible than glass or many other building materials.

**Electrical/Thermal Properties:** Acrylic sheet is an excellent insulator. Its surface resistively is higher than that of most plastics. Continuous outdoor exposure has little effect on its electrical properties. It is also a better thermal insulator than glass. Acrylic Sheet is often used as the sidewall of LTD Stirling engine pressure chambers because its low thermal conductance characteristics make it a good thermal break between the warm and cold sides of the engine.

**Chemical Resistance:** Acrylic sheet has excellent resistance to attack by many chemicals. It is affected, in varying degrees, by benzene, toluene, carbon tetrachloride, ethyl and methyl alcohol, lacquer thinners, ethers, ketones and esters. Acrylic sheet is not affected by most foods, nor are foods affected by it.

**Light Transmission:** Colorless acrylic sheet has a light transmittance of 92%, which is greater than glass. Welds can be made to be almost invisible, and rough surfaces can be polished so the finished parts are brighter and clearer than glass.

**Fire Precaution:** All brands of acrylic sheet are a combustible thermoplastic. The self-ignition temperature range is 830-910°. Protect it from flames and high heat.

**Cleaning:** Wash acrylic sheet with a mild soap or detergent and plenty of lukewarm water. Use a clean soft cloth, applying only light pressure. Rinse with clear water and dry by blotting with a damp cloth or chamois. Grease, oil or tar may be removed with a good grade of hexane, aliphatic naphtha, or kerosene. These solvents may be obtained at a paint or hardware store and should be used in accordance with manufacturers' recommendations.

Any oily film left behind by solvents should be removed immediately by washing.

**Do Not Use:** Window cleaning sprays, kitchen cleansers, gasoline, benzene, carbon tetrachloride or lacquer thinner.

Static electricity can attract dust to acrylic sheet. To reduce it, use an anti-static cleaner which is available from your ACRYLITE dealer, or consider using a de-ionizing air gun.

**Masking:** Acrylic sheet comes covered on both sides with a masking of latex paper, polyethylene, or vinyl film. The masking protects the sheet from scratching during storage and handling. ***Be sure to leave the masking in place during most phases of fabrication and installation.*** Except for intricate detail work, you should remove the masking only when your project is completed.

You can remove the masking paper from larger flat areas with a cardboard tube—rolling the paper around it. All masked acrylic sheet should be kept away from heat and sunlight, and masking should be removed soon after installation. If the adhesive has hardened, moistening the paper with aliphatic naphtha, hexane, or kerosene will help soften it. Never use a knife or scraper to remove masking.

### *Do's*

- Keep masking on as long as possible through your fabrication operations. Always remove masking from internal parts prior to assembly.
- ***Wear safety glasses when working with power tools.***
- ***Wear gloves when handling large sheets to prevent cuts.***
- Use drill bits which are designed or reground for acrylics and carbide tipped circular saw blades and router bits.
- Make sure all your tools are sharp.
- Use water as a coolant when cutting sheets over 1/4" (6.0 mm) thick or drilling sheets over 3/16"(4.5 mm) thick.
- Use plenty of water when cleaning acrylic sheet to help prevent scratching.

### *Don'ts*

- Don't store acrylic sheet near radiators or steam pipes or in direct sunlight.
- Don't leave masking on if it will be exposed to the outdoors (sun or rain).
- Don't mark with a punch marker.
- Don't use saw blades that have side-set teeth. Saw teeth should be carbide tipped with 0°-15° rake and slight radial clearance.
- Don't bring the material in direct contact with heaters.
- Don't subject the sheet to high surface temperatures during polishing.
- Don't use glass cleaning sprays, scouring compounds or solvents like acetone, gasoline, benzene, carbon tetrachloride, or lacquer thinner.
- Don't heat acrylic sheet in a kitchen oven.

### *Cutting Acrylic*

Acrylic sheet can be sectioned in a wide variety of ways, with either hand tools or power tools. The method you choose will likely depend on the particular tools available to you, but all tools cannot be used in all cases. Your choice of tool and techniques should be based on the type of acrylic sheet used, the thickness of the sheet, and the shape of the particular cut. This section, though not comprehensive, gives some guidelines for choosing the right tool, and using it properly to get the best results when cutting acrylic sheet.

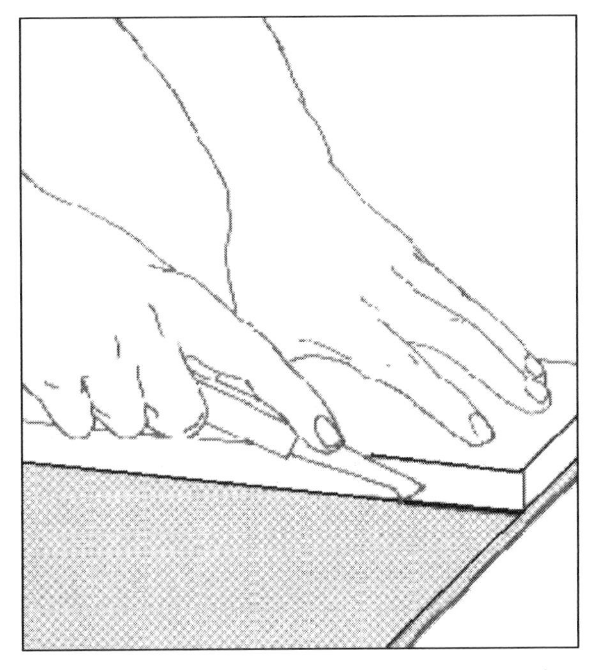

**Figure 9 - Always draw the scribing knife along a straight edge.**

#### Cutting with a Knife or Scriber

Acrylic sheet up to 3/16" (4.5 mm) thick may be cut by a method similar to that used for cutting window glass. Use a scriber of some kind—a scribing knife, a metal scriber, an awl, or even a sturdy craft knife—to score the sheet. Draw the scriber several times [7 or 8 times for a 3/16" (4.5 mm) thick piece] along a straight edge held firmly in place. It is best not to remove the protective masking. Make the cuts carefully using firm, even pressure. For best results, make each stroke cleanly off the edge of the sheet.

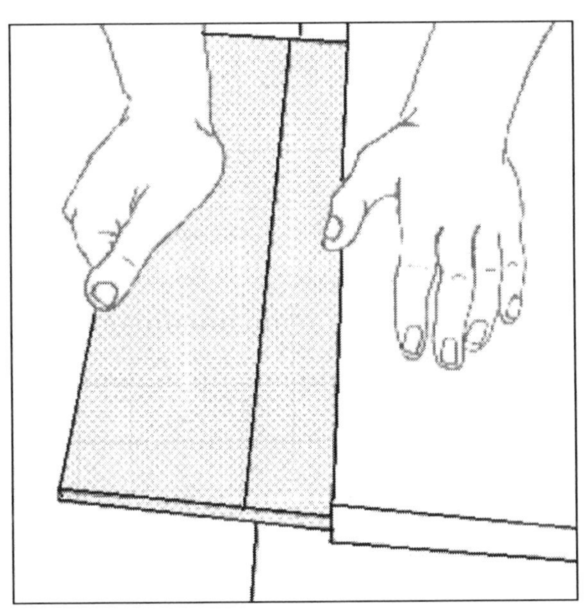

**Figure 10 - Break sheet over edge of table after scribing.**

Be very careful that the blade of your scribing tool does not wander away from the straight edge when making repeated strokes. It may help to clamp the straight edge in place before making the first cut. Make additional cuts slowly and deliberately so that the blade does not wander and create multiple break lines.

Then, clamp the acrylic sheet or hold it rigidly under a straight edge with the scribe mark hanging just over the edge of a table. Protect your hands with a cloth, and apply a sharp downward pressure to the other side of the sheet. It will break along the scratch. Scrape the edges to smooth any sharp corners. This method is not recommended for long breaks or thick material.

### *Cutting with Power Saws*

**CAUTION! Wear safety glasses when working with power tools.**

With any type saw, blades should be sharp, and free from nicks and burrs. Special blades for cutting acrylics are available for most types of saws. Your authorized ACRYLITE acrylic sheet distributor should have them in stock. (Look in the yellow pages for “Plastics”.) Otherwise, use carbide-tipped blades designed for cutting plastics available at industrial products suppliers like Sears. Teeth should be fine, of the same height, evenly spaced, with little or no set.

**Table saws and circular handsaws:** Use hollow ground, high-speed blades with no set, and at least 5 teeth per inch (25 mm), such as those used to cut copper and aluminum. If you intend to do a lot of cutting use carbide-tipped blades designed for plastics (a triple chip type tooth design is recommended). These give a cleaner cut in acrylic sheet. Set the blade to project approximately 1/8" (3 mm) above the surface of the sheet being cut. This will reduce edge chipping.

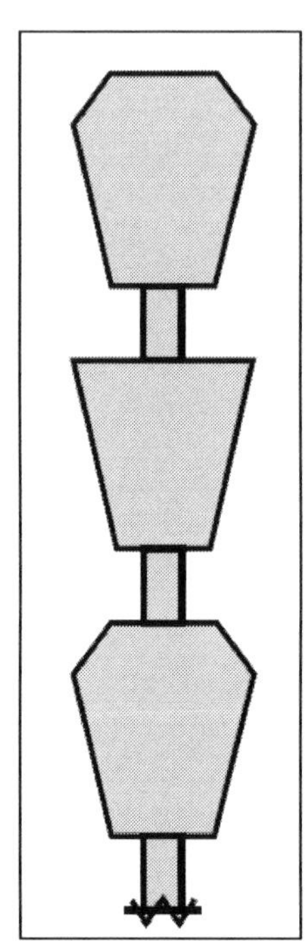

**Figure 11 - Triple-chip alternate tooth design blades run cooler and work well in plastics.**

A 10-inch diameter, 80-tooth blade is recommended for all-purpose cutting on a table saw. The triple-chip design will have one cutter that cleans the center of the cut followed by another cutter that cleans the corner of the cut. The radial clearance allows the chips to clear, and reduces friction and heat while cutting.

When cutting with a hand-held circular saw, clamp the sheet securely to the work surface to minimize vibration. A wood block 1" x 3" (25 x 75 mm) clamped on top of the sheet spreads the clamping force and can act as a guide for the saw.

No matter which type of saw you use, the sheet must be held firmly and fed slowly and smoothly to prevent chipping. Be sure the saw is up to full speed before beginning to cut. Water-cooling the blade is suggested for thicknesses over 1/4" (6 mm), especially when edge cementing will be performed. (Only use water cooling if your saw is equipped for water cooled cutting.)

I did not use a table saw for any acrylic cutting, nor do I recommend it. The parts needed for these engines are very small, and the potential for harm when using a table saw with acrylic is too high for my comfort level. (That, and the fact that I did not want to shell out the extra money for the specialized blade.) ***Do not attempt to cut acrylic with a circular saw if you are not using a blade designed specifically for the task.***

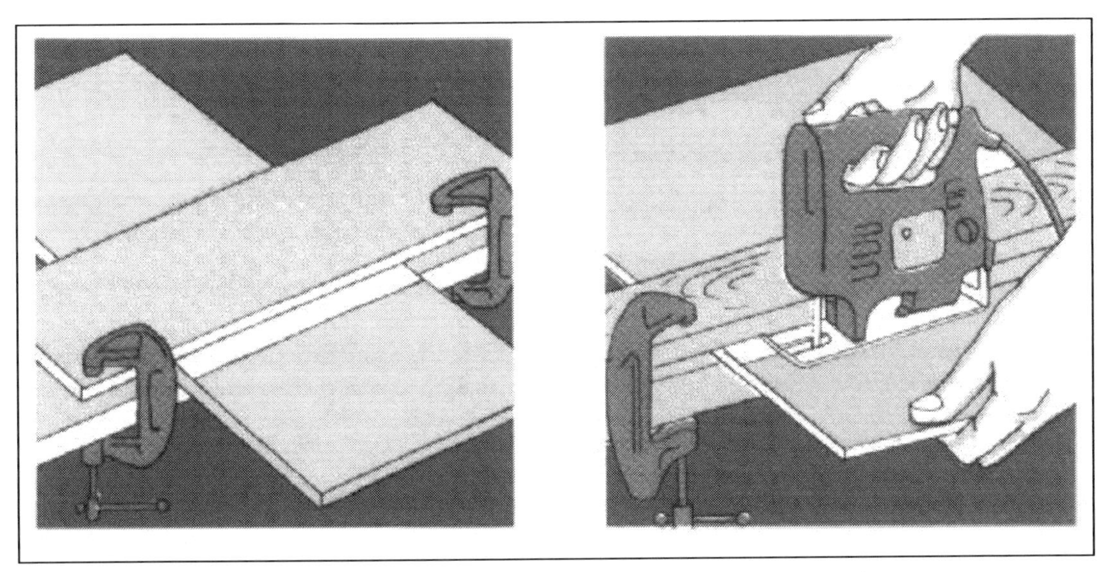

**Figure 12 - Clamp material well before cutting and use the clamping board as a saw guide.**

**Saber saws:** Use blades which have a slight set, such as the blades recommended for cutting metals or other plastics. Be sure they are sharp. The blades you use for cutting acrylic should never be used to cut other materials. Set them aside. Use them only for acrylic sheet.

High speed is best for cutting acrylic sheet with a saber saw. Always be sure the saw is at full speed before beginning to cut. Press the saw shoe firmly against the material, and don't feed too fast. Water cooling is suggested for cutting acrylic sheet over 1/4" (6 mm) thick, but only on saws where water cooling is a safe and acceptable option.

**Band saws or jig saws:** Band saws and jig saws are excellent tools for cutting acrylic sheet. But because of their relatively thin blades, they are not recommended for cutting acrylic sheet over 1/4" (6 mm) thick. Use blades with a slight set, and about 10 teeth per inch (25 mm). Saw blade manufacturers recommend a blade that will always have at least 3 teeth in contact with the material being cut. Feed acrylic sheet at a rate 10 times faster than you would feed steel. Blades may break easily in acrylic, so operate accordingly.

If the material gets too warm during the cutting process the acrylic will begin to melt and heal the cut behind the saw blade, and can actually trap the saw blade in the middle of the piece being cut. So, to avoid starting a collection of saw blades welded into the middle of acrylic sheets, always attempt a test cut using scrap material whenever possible.

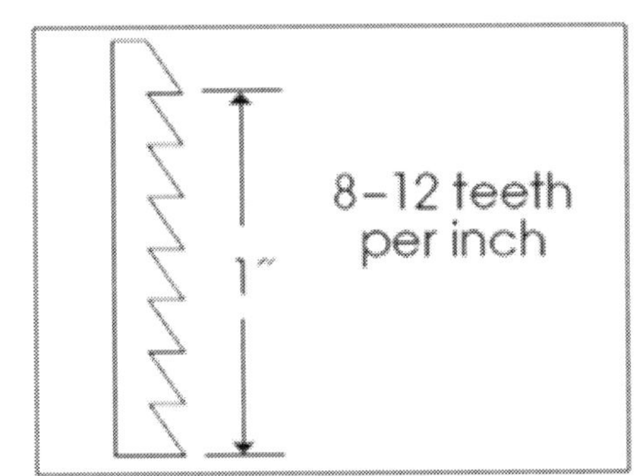


**Figure 13 - Band saw blade design.**

I have had reasonable success using a jig saw if I lubricate the blade with light oil or silicone before each cut. This helps it run cooler by reducing friction and the associated heat buildup. I have also had good luck using air cooling by blowing compressed air at the blade during the cutting process.

I did not have access to a band saw at the beginning of this project, but I acquired one in time to make a few parts for the last engine. I had the best results with the band saw for cutting acrylic sheet. I did not have to use any special lubrication or cooling techniques and it was very easy to cut a straight line. A simple circle cutting jig on a band saw would make it very easy to create acrylic flywheels too.

### *Cutting with hand saws*

Acrylic sheet may be cut with almost any type of hand saw. And while good results are possible with hand saws, the techniques involved are considerably more difficult than with power saws. Practice on scrap material before attempting to make critical cuts.

With any hand saw, it is most important that the blades be kept sharp. For best results, the teeth should be of uniform size and shape, and have very little set. Every effort should be made to prevent vibration or stress while cutting. Flexing at the point of the cut or binding of the saw blade may cause the acrylic to crack. Clamp the material securely. Keep the saw straight when cutting, and apply very little pressure.

Let the blade do the work. With practice and proper care, you can get good results.

**Straight saws:** Straight saws or cross-cut saws may be used for long, straight cuts on acrylic sheet of almost any thickness. The saw should have a hollow-ground blade with very little set and at least 10 teeth per inch (25 mm). Make certain the material is firmly clamped and supported. Hold the saw at an angle of about 45° from vertical, and be sure to keep it straight.

**Coping saws:** Coping saws or scroll saws are good for shorter cuts, curved cuts, or even intricate designs. Use very narrow blades with only a slight set.

**Hacksaws:** These hand saws for cutting metal may also be used for short cuts in acrylic sheet. Choose a blade with approximately 18 teeth per inch (25 mm). Use a smooth, even stroke. Apply very little pressure.

### *Routing and shaping*

We don't do any routing or shaping for this project. Acrylic sheet, however, can be machined with standard woodworking routers, in much the same way as wood. You'll find many uses for portable hand routers and small table routers. Use them to cut patterns into edges, or large holes out of pieces of acrylic sheet. For best results, use single-fluted bits for inside circle routing, and double-fluted bits for edge routing.

Routers are designed to operate at high speeds. 10,000 to 20,000 rpm is recommended for acrylic sheet. And because routing speeds are so high, vibration must be scrupulously avoided. Even small vibrations can cause crazing and fractures in acrylic sheet during routing.

### *Turning*

Turning is the only practical way to produce most round cross-sectioned parts such as knobs, furniture legs, and vases. Acrylic sheet can be turned on almost any type lathe. Bits designed especially for cutting acrylic are available. If you own a wood lathe you could use it to make the flywheels used in the first two engines in this book.

### *Drilling Acrylic*

Any kind of hand or power drill may be used for drilling acrylic sheet. A stationary drill press is the preferred tool because it gives better control and greater accuracy. But a drill press won't be applicable in all instances, and with a little care, proper technique, and a correctly-ground drill bit, you can get good results with an ordinary hand drill.

For best results, use drill bits designed specifically for acrylics, available in many tool stores, and stores where acrylic sheet material is sold.

Regular twist drills can be used, but the cutting edges must be modified to prevent the blade from grabbing and fracturing the plastic. Acrylic sheet is relatively soft. Your drill should have an edge that cuts with a scraping action. To obtain this you can modify your drill bit by grinding small "flats" onto both cutting edges with a medium or fine-grit grinding wheel, or a pocket stone. The flats can be parallel to the length of the drill and about 1/32" (1 mm) wide. Tip angle should be between 60° and 90°.

For the best possible finish inside the hole, use a drill with smooth, polished, slow-spiral flutes which will clear the hole of all shavings without marring or melting the walls.

Two continuous spiral chips or ribbons will emerge from the hole if the drill is correctly sharpened and operated at proper speed.

**Figure 14 - The carbide tips on these simple drill bits are designed for cutting material such as acrylic. Sets like this one are available for between $5 and $10 at discount tool stores. These came from Harbor Freight® Tools.**

When drilling a hole three times deeper than the diameter of the drill, a lubricant or coolant should be used. This will help remove chips, dissipate heat, and improve the finish of the hole. Rough, irregular, or fuzzy holes can lead to cracking and breaking months after the piece has been completed.

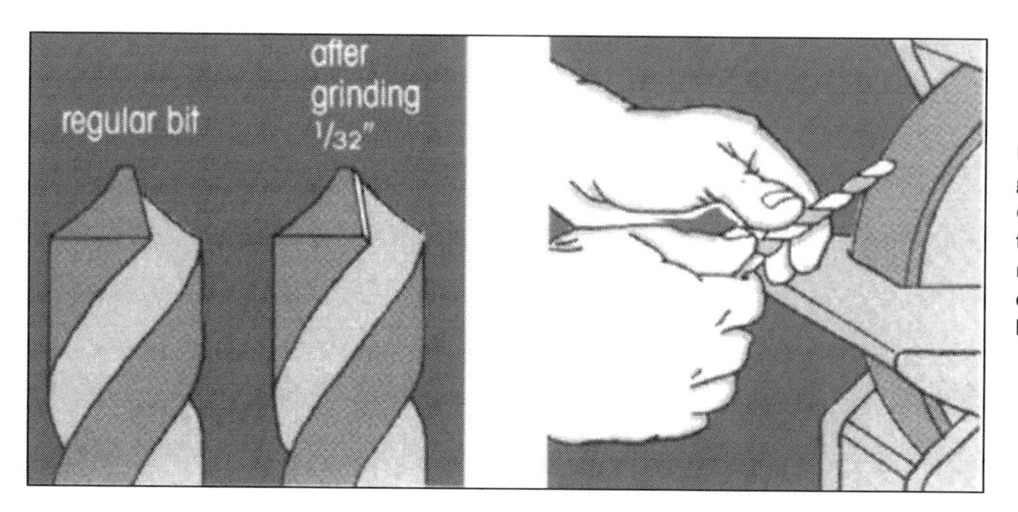


**Figure 15 - Small flats ground into the cutting edge will prevent the drill from grabbing. This modification makes the drill work poorly in wood, however.**

### *Forming Acrylic*

Acrylic sheet becomes soft and pliable when heated—behaving almost like a sheet of flexible rubber. It may then be formed into almost any shape. As the sheet cools, it hardens and retains the formed shape, provided it has been held in place during the cooling process.

Every formulation of acrylic sheet will have a slightly different temperature at which it becomes pliable. For most brands, this is at about 300° Fahrenheit.

Do not exceed the safe working temperature for the material by continuing to increase the temperature beyond the point where it becomes pliable. Excessively high temperatures may cause the sheets to blister and burn. **Never heat acrylic sheet in a kitchen oven.** Acrylic sheet gives off highly flammable fumes when decomposed by overheating. These gases are potentially explosive if allowed to collect in an unventilated area.

Most kitchens ovens do not have accurate temperature control. Temperatures can be off as much as 75° (42°C), possibly allowing the acrylic to overheat.

And because air is not forcibly circulated in a standard kitchen oven, the fumes will accumulate. When they come into contact with the heat source, there is likely to be an explosion. Repeat: ***Do not*** heat acrylic in a kitchen oven.

### *Forming With a Strip Heater*

A strip heater is without doubt the most useful acrylic-forming device in the home craftsman's arsenal. If you get hooked on working with acrylic you will eventually invest in one. While it is not a requirement for any of the projects in this book, knowing how they work could be useful.

Forming jigs and clamps should be used for best results. They can be made of wood and used over and over. Make preformed jigs for certain angles, or even special shapes for individual projects. Variable angle jigs can be made with two pieces of wood hinged together and held at the desired angle with a variable brace. Felt, flannel, or flocked rubber should be used to line any surfaces that may come into contact with the heated acrylic. Wear heavy cotton gloves when handling heated acrylic sheet. They'll protect your hands, as well as the sheet.

Polycarbonate sheet looks and feels just like acrylic sheet, but it is a harder material that does not work well for thermoforming applications.

We only make a few small bends for the projects in this book. The bending we have to do can be accomplished with a heat gun, or a very hot hair dryer.

### *Other Forming Techniques*

Acrylic sheet may be formed into almost any shape. However, specialized heating and forming equipment is usually required for all but the simplest projects. Furthermore, while many of the forms and jigs required for two and three dimensional forming can be easily made out of wood in the home shop, such projects are beyond the scope of this book. Many excellent books are available covering all types of acrylic forming. They deal with techniques such as drape forming, plug and ring forming, surface molding, blow and vacuum forming, and even design, construction, and use of specialized ovens for heating acrylic sheet.

## Joining Acrylic

Acrylic sheet can be joined with solvent cements to form strong, durable, transparent joints. Some formulations of acrylic sheet have added coatings for improved resistance to scratching or for filtering ultraviolet light. If solvent cementing on or to one of these coated surfaces is necessary, the coating must first be removed by wet sanding with 500 grit, or finer, sandpaper. Most acrylic sheeting will not have these coatings.

**The ingredients in most solvent cements are hazardous materials, and extreme care should be observed using proper ventilation and handling techniques as recommended by the manufacturer of these products. Always follow the manufacturer's recommendations and instructions when using these and any other products.**

The ultimate strength and appearance of your joints will depend on how carefully you make them. Getting really good joints requires a lot of care, and considerable skill. Practice on scrap pieces. The more experience you have, the better your work will be.

**Observe some basic precautions when working with acrylic solvents:**

- Always work in a well-ventilated area.
- Do not smoke—solvents are highly volatile and flammable.
- Protect skin from contact with cement.
- Do not attempt to cement acrylic sheet in temperatures under 60°F (15°C). Temperatures from 70° to 75°F (21° to 24°C) are ideal.
- Always follow the cement manufacturer's recommendations.

### *Preparation of the joint*

All surfaces that are to be joined should fit together accurately without having to be forced. Flat, straight surfaces are easiest to work with. Any area that is part of the original surface of the sheet should be left untouched.

A smooth cut made with a cooled power saw also should be left alone. However, if the area to be joined has a saw cut that is rough, it should be wet sanded or finished with a router to get a flat, square edge. Do not polish edges that are to be cemented. Polishing leaves a highly stressed, convex edge with rounded corners. It will make a very poor joint. Always remove the masking from around the area to be joined.

### *Capillary cementing*

Capillary cementing is probably the most popular method of joining acrylic sheet. It works because of the ability of low-viscosity solvent-type cement to flow through a joint area by capillary action. Properly done, it yields strong, perfectly transparent joints; however, it won't work at all if the parts do not fit together perfectly.

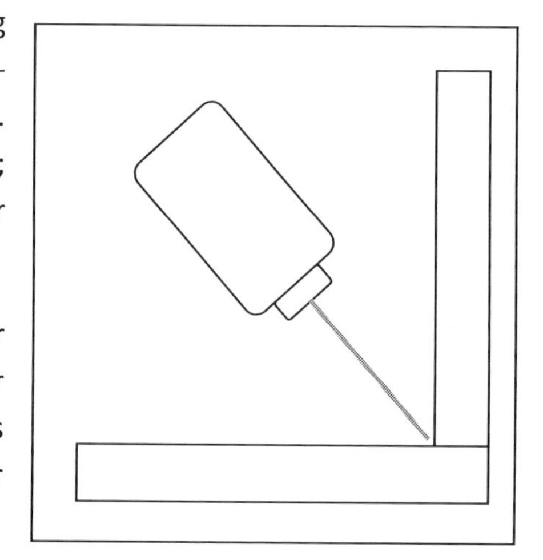

**Figure 16 - The joint should fit snug before applying cement. Keep the joint flat so cement does not run out.**

Solvent cements are available from your acrylic sheet distributor who can recommend the ones that are best for your particular projects. (If the price for the solvent and solvent applicator is beyond your budget for this project, clear epoxy glue makes a fair substitute for solvent joints.)

First make sure the parts fit together properly. Then hold the pieces together using a jig that will support the pieces firmly but will permit slight movement as the joint dries. It is important that the joint be kept in a horizontal plane, or the cement will run out of the joint.

Apply the cement carefully along the entire joint. Apply it from the inside edge, whenever possible on a box-corner type joint, and from both sides, if possible, on a flat piece. A special needle-nozzled applicator bottle available from your acrylic sheet distributor is recommended, and will make the process much easier for you.

If the cement does not flow completely into the joint, try tilting the vertical piece very slightly (about 1°) towards the outside. This should allow the solvent to flow freely into the entire joint. Always let the joint dry thoroughly (usually 10-30 min.) before moving the part. Maximum bond strength will not be reached for 24 to 48 hours.

### *Dip or soak cementing*

This method of cementing acrylic sheet involves dipping the edge of one of the pieces to be joined directly into the solvent. It is very important that only the very edge be dipped. Exposing too much area to the solvent will result in a weak, slow-setting joint.

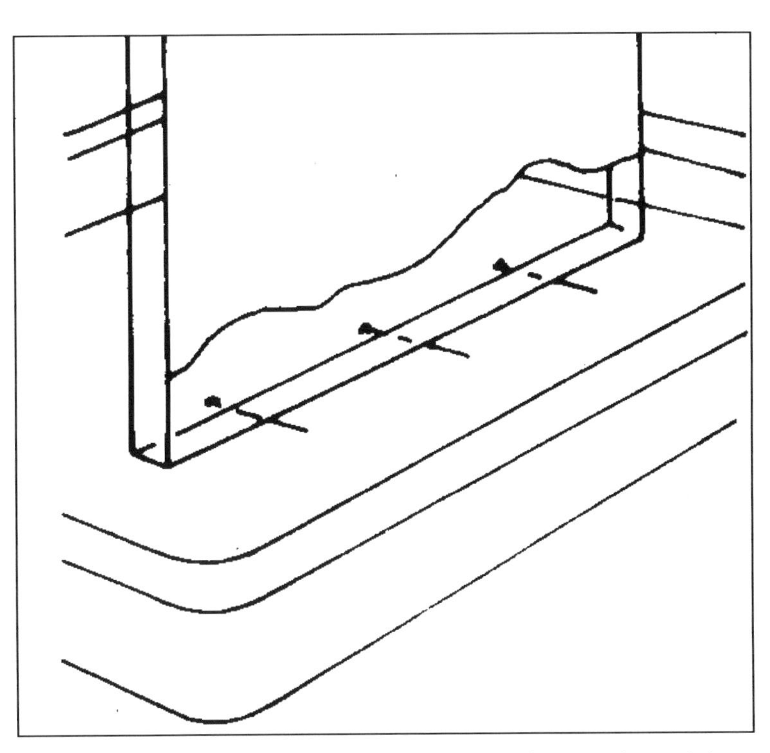

**Figure 17 - Soak Cementing is an alternative method for gluing longer joints. Material must be supported on pins or wire brads.**

You will need a shallow tray in which to dip the acrylic. The tray can be made of aluminum, stainless steel, galvanized steel, or glass. Do not use plastic— the solvent may dissolve it.

Place short pieces of wire, pins, or brads into the tray to keep the edge of the acrylic sheet from touching the bottom of the tray. The tray must be level. Pour solvent cement into the tray so that it just covers all the brads—and covers them evenly.

Now carefully place the edge to be cemented into the tray so that it rests on the brads. You can hold the piece upright by hand, but it is better to use some kind of support to hold the piece in place while it soaks. A couple of padded clamps attached to the sheet, and resting on the edge of the tray are fine. Heavy pieces of wood placed against each side of the sheet will also work. Slotted wooden supports are usually used for production work, but anything that will hold the piece firmly upright is sufficient.

The acrylic sheet should be left in the solvent from 1/2 to 2 minutes, depending on the thickness of the sheet, the type of solvent used, and the bond strength required. Soaking time should be long enough to allow the edge of the sheet to swell into a "cushion." As soon as an adequate cushion is formed, the piece must be removed. Hold it for a few seconds at a slight angle to allow the excess solvent to drain off. Then carefully, but quickly, place the soaked edge precisely into place on the other part to be joined. Hold the parts together for about 30 seconds without applying any pressure. This will allow the solvent to work on the surface of the other piece.

After 30 seconds you can apply some pressure to squeeze out any air bubbles. But be very careful not to squeeze out the cement.

When the pieces are joined, the part should be placed in a jig to maintain firm contact for 10 to 30 minutes. Do not allow the parts to move during this critical time. Allow the joint to set for another 8 to 24 hours before doing any further work on it.

### *Viscous cementing*

Viscous cements are used to cement joints that can't be easily cemented by capillary or solvent soak methods—either because they are difficult to reach, or because the parts don't fit properly together. Viscous cement is thick. It will fill small gaps, and can make strong, transparent joints where solvent cements can't. Viscous cements are available from your acrylic sheet distributor. Or you can make your own viscous cement by dissolving chips of clear acrylic sheet in a small amount of solvent. Let the solution stand overnight.

Apply the cement like glue with a brush, spatula, or directly from the applicator tube.

Remove the masking material from around the joint area, and carefully apply a small bead of cement to one side of the joint. Then gently join the pieces as described under "Soak Cementing."

Masking tape may be applied to protect the area around the joint, but it should be removed carefully after about 5 minutes, while the cement is still wet. Don't touch the parts at all for the first critical 3 minutes, or the joint will not hold. The part may be carefully moved after 30 minutes, but don't do any additional work on it for 12 to 24 hours.

## Finishing Acrylic

### *Scraping*

Many of the techniques used to cut acrylic sheet can leave a rough edge that is usually unsuitable either as a finished edge, or to join to another piece of acrylic. It is necessary to smooth and square the edge of the sheet. You can do this by a number of different techniques, depending on the finish desired.

The first step, and perhaps easiest technique, is scraping. A scraper can be almost any piece of metal with a sharp, flat edge. The back of a hacksaw blade, the back of a knife blade, or a tool steel blank are ideal. Special acrylic scraping tools are also available if one has the interest. Whatever tool you use must have a sharp, square edge.

### *Filing*

It is easy to file acrylic sheet to a surface ready for final polishing. The filing, however, must be done correctly, and carefully.

Almost any commercial file can be used. But the quality of the finish will depend on your choice of file coarseness. A 10 to 12 inch (250 to 300 mm) smooth-cut file is recommended for filing edges, and removing tool marks. Other files—half round, rat tail, triangular files, and even small jewelers' files are good for smoothing insides of holes, cutting grooves and notches, or finishing detail.

File in only one direction. Keep the teeth flat on the surface of the acrylic sheet, but let the file slide at an angle to prevent the teeth from cutting unwanted grooves in your work.

Always keep your files clean and sharp. Wire brush them often to prevent the teeth from filling up. And don't use your acrylic files for working metal or other materials that might dull the teeth.

For small work, try clamping the file in a vise, and rubbing your work across the file.

### *Sanding*

Before acrylic sheet is ready to be polished, it should be sanded to a smooth, satiny finish. As with filing, the quality of the final finish will depend on the grades of sandpaper used. The finer the final grit, the smoother the finish will be. It will usually take at least three steps to get a good finish.

First, if there are scratches deep enough to require it, start with coarse grit No. 120 sandpaper. Use it wet. When the original scratches are completely removed, sand with a medium grit paper—220 is good—to remove the scratches from the coarse paper. Use the medium grit paper wet as well. Finally, sand to a satiny finish with a fine grit, wet-or-dry No. 400 paper. Fine grit paper should always be used wet to keep the paper from clogging and obtain a smoother finish. Rinse the paper frequently. Grits as fine as 600 may be used.

Always wipe your work clean when changing to a finer grit. Be sure all deep scratches have been removed.

**Sanding by hand:** Hand sanding acrylic sheet is very much like hand sanding wood. Most of the same techniques apply. But sanding acrylic must be done with far greater care. You should always use a wooden or rubber sanding block. When removing scratches, be sure to sand an area that is slightly larger than the scratch. This will help prevent low spots. Sand with a circular motion. Use light pressure and plenty of water with wet-or-dry papers.

As you get the feel of working with acrylic sheet, your own observations and experience will be your best guide to determining how coarse a grade to start with on each particular job, and how many different grades will be needed to do the job most efficiently. Don't be afraid to experiment with different sanding techniques and different types of blocks. You'll learn a lot of new tricks—perhaps the very one you'll need to help solve your next problem.

**Sanding with power sanders:** Almost any commercial power sander can be used to work acrylic sheet. Naturally, different types of sanders are preferred for different operations. As a basic rule, use them as you do when sanding wood. They should, however, be operated with lower pressure, and at slower speeds. Experiment on scrap pieces. All wet-or-dry machine sanding should be done wet—especially with grit sizes of 150 or finer.

### *Polishing*

The original high luster of acrylic sheet can be restored to the edges and surfaces by polishing with a power driven buffer. It is also possible to polish acrylic sheet by hand using a soft cloth and a very fine abrasive. But hand buffing is an extremely tedious process. You're likely to get a sore arm long before you get a finely polished surface.

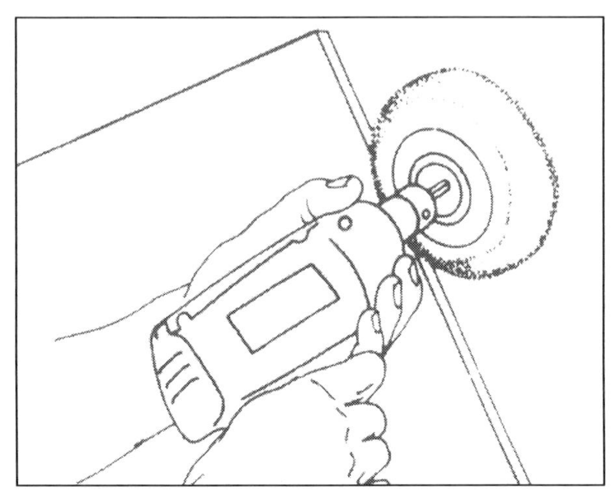

**Figure 18 - Edge polishing with a hand drill.**

Power-driven buffing tools are recommended almost without exception. Because inexpensive buffing wheels are available as an attachment for any electric drill, equipment should not be a problem. Buffing wheels and compounds good for acrylics are sold by specialty stores and acrylic sheet distributors, but special wheels are not really necessary. A good buffing wheel for acrylic sheet will consist of layers of 3/16" (4.5 mm) carbonized felt, or layers of unbleached muslin laid together to form a wheel between 1 and 3 inches (25 and 75 mm) thick. The larger the wheel, the better. But don't use one too large for your equipment. The wheel should reach a surface speed of at least 1200 feet per minute (370 m per minute). Speeds up to 4000 feet per minute (1220 m per minute) are useful for acrylics.

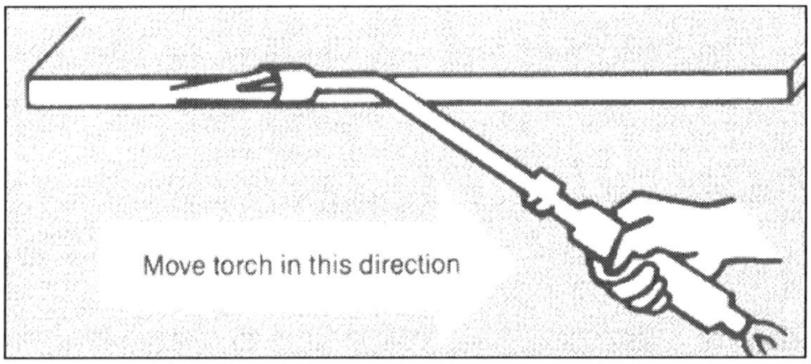

Solidly stitched wheels with rows of concentric stitching should be avoided. They are often too hard and may burn the acrylic. Never use a wheel at speeds higher than its rpm rating. Never use a wheel that has been used to polish metal. Traces of the metal may remain to scratch the acrylic sheet.

**Figure 19 - Flame polishing is another quick method for finishing an edge. Do not polish edges that will be glued.**

Acrylic sheet should be polished using a commercial buffing compound of the type used for polishing softer metals such as silver or brass. Or you can use a non-silicone car polish that has no cleaning solvents in it.

First, however, tallow should be applied to the wheel as a base for the buffing compound. Just touch the tallow stick to the spinning wheel. Then quickly apply buffing compound.

To polish, move the piece back and forth across the wheel until you get a smooth, even polish. Be careful not to apply too much pressure, and keep the work constantly moving across the wheel. This will help prevent heat buildup which can mar the surface by melting or smearing it.

For safety reasons, it is important not to start polishing near the top of the sheet. The wheel may easily catch the top edge, tearing the piece of acrylic sheet out of your hands and throwing it across the room . . . or at you. Always wear safety glasses and be extremely careful.

Begin polishing approximately one-third of the way down the sheet, and keep moving it back and forth until you've reached the bottom edge. Then turn the sheet around and repeat the process on the other half.

A propane torch can also be used for polishing the edge of acrylic sheet material. Be careful that you do not get the material too hot when using a torch because it will burn if overheated.

### *Some final words on working with acrylic*

Any material that is cut with a power tool presents a certain amount of risk to the user. Always use safety precautions and safety equipment when using any tool.

**Important notice:** The information and statements herein are believed to be reliable but are not to be construed as a warranty or representation for which we assume legal responsibility. Users should undertake sufficient verification and testing to determine the suitability for their own particular purpose of any information or products referred to herein. No warranty of fitness for a particular purpose is made.

Nothing herein is to be taken as permission, inducement or recommendation to practice any patented invention without a license.

My thanks to CYRO Industries, manufacturers of ***ACRYLITE® acrylic sheet*** products, for providing a wealth of information about how to work with acrylic. Visit their website for more information about working with acrylic. There you will find detailed information about tool selection and advanced techniques for cutting, forming, and finishing acrylic sheet products.

# Chapter 7: Thermoforming Vinyl

### *What is Thermoforming?*

Thermoforming is a process where a sheet of plastic or vinyl film is heated to a point where it becomes flexible and can be formed or shaped over a mold. Vacuum is applied to hold the film against the form until it cools. After it cools, it will retain the new shape provided by the mold.

### *Benefits of Thermoforming*

The engines described in this manual use a vinyl diaphragm rather than a drive piston as the main drive mechanism. Thermoforming the diaphragm will reduce friction in the drive assembly and will improve efficiency by capturing energy that would otherwise be lost. Forming the diaphragm so that it is a perfect fit for the drive cylinder will remove any wrinkles that are the cause of pressure leaks.

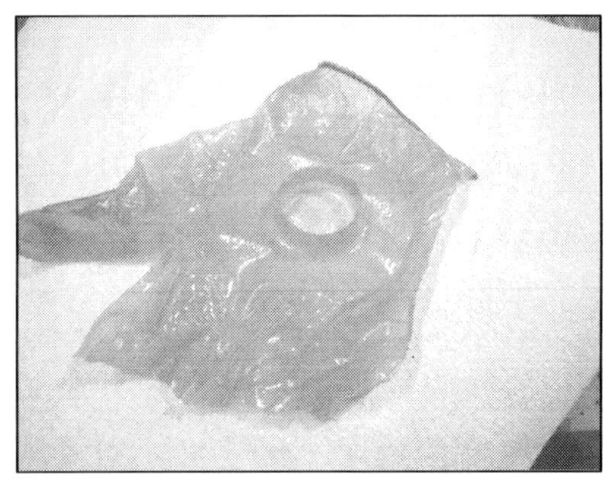

**Figure 20 - A piece of vinyl glove has been thermoformed to create a diaphragm.**

Once again, let's look at the famous pop can engine. It uses a balloon diaphragm to drive the motor. (Do a quick search on YouTube for "Pop Can Stirling Engine") As you watch the engine operate you can see there is a lot of motion in the balloon that is not being put to use to drive the motor. The motor uses energy to stretch the balloon that could be used to turn the flywheel.

You may also notice there are wrinkles in many diaphragms. Some wrinkling cannot be avoided, but excessive wrinkles caused by excess fabric may slow the engine because it is not effectively harnessing all the available energy.

With the latex balloon, there is always a question of how tight it should be stretched. Some designers tell you to stretch it taught, others will say to leave it loose. A tight diaphragm consumes energy because the engine must stretch the diaphragm to get it to move off its neutral position. If the diaphragm is left too loose, expanding and contracting gasses are not fully utilized and the effort is wasted.

Thermoforming allows us to create a diaphragm that has the following characteristics:

1. It is shaped to fit the drive cylinder.
2. It eliminates many of the wrinkles that decrease efficiency in flat diaphragms.
3. It has less wasted motion compared to latex balloon diaphragms.
4. The material is flexible enough to allow motion yet has very little energy lost to stretching.
5. Small amounts of slack in the shaped diaphragm do not create significant energy loss.

The more high-tech Stirling engine designs use a drive piston in a cylinder. It is also one of the most difficult parts to make for any Stirling engine. A graphite piston moving in a glass cylinder is a very effective and beautiful drive mechanism, but building one requires incredible precision. The piston has to fit the cylinder tight enough so that air cannot escape, yet loose enough so that it can slide with no (or very minute) friction.

In my experience as a low-tech Stirling engine builder I have had some success with drive mechanisms made from rubber balloons, pistons and cylinders made from polished brass tube, and from thermoformed vinyl. The thermoformed vinyl has been the most effective mechanism in all of my home built engines.

I have searched online for a manufacturer that might sell the graphite and glass drive assemblies. I finally found one website that offered the parts for sale, but they appear to be permanently out of stock.

Thermoforming is not difficult, and the materials needed are not expensive. It may take a little practice to get the process to work, but once it does, you will probably make yourself a small collection of diaphragms to experiment with and keep on hand as spares.

## Equipment Needed for Thermoforming

The equipment list is quite short.

- Vacuum Cleaner
- Scrap of flat wood
- Heat Gun
- Vinyl
- Frame
- Form

Don't be intimidated by anything in this list. You should have most of what you need laying around with your craft supplies and tools. The only uncommon tool used in this process is the heat gun, and if you are lucky, you might find you have a hair dryer that gets hot enough to do the work.

I am assuming most of us already have a **vacuum cleaner**. I used my shop vacuum for this, and it was almost too powerful for the job. All you really need for this is a vacuum cleaner that has a hose attachment. A household vacuum will do the job.

The **scrap of flat wood** will become our makeshift vacuum table. As you can see, my vacuum table is a 6" x 15" piece of wood that I pulled from the scrap bin. I traced the outline of the end of my vacuum hose and then drilled a collection of 1/8" holes within the circle. I cover the vacuum table with a paper towel before using. That creates a smooth porous work surface. These makeshift vacuum tables are so quick and easy to put together that I have been known to make a new one from a scrap of wood rather than dig through the garage to find the old one!

**Figure 21 - A scrap of wood, a vacuum cleaner, and a couple of bungee cords are all it takes to make a very effective vacuum table for thermoforming small parts.**

The **heat gun** is used to warm the vinyl. I bought one on sale at Harbor Freight Tools for under $10 and it has proven to be a useful tool. It is almost too hot for this application. You must take care that you do not overheat your vinyl or it will melt or burn. There are probably some creative alternatives for those who can't get access to a heat gun. Some hair dryers will get hot enough to work for this process.

The **vinyl** comes from disposable vinyl gloves available from paint and hardware stores. I have found they come in at least two grades. The heavy vinyl will provide a more durable diaphragm when thermoformed. The lightweight gloves are more prone to tearing during the thermoforming process and are much thinner after forming. Each grade has advantages but I prefer the heavier material. They are cheap, so I have both.

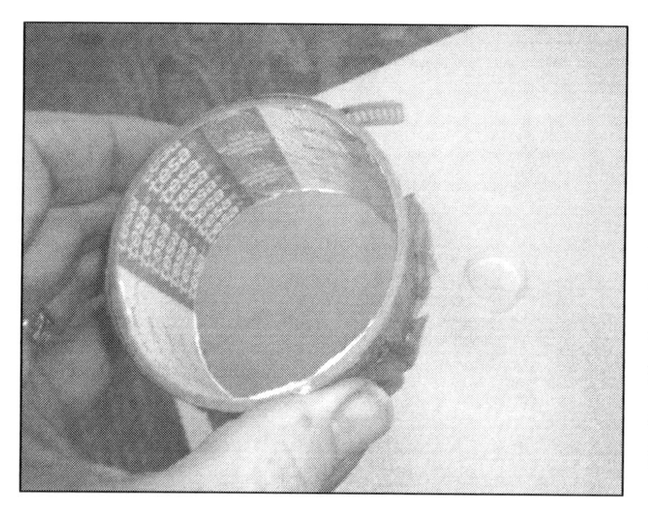

**Figure 22 - An old masking tape roll is used as a hoop frame to hold the vinyl sheet as it is heated.**

**Note: Latex rubber gloves and Nitrile rubber gloves will not work for thermoforming. They must be vinyl.**

The **frame** is simply a method to hold the vinyl flat while it is heated and stretched over the form. I found that an old masking tape roll and a rubber band provided excellent results. The vinyl sheet is stretched across the opening of the old tape roll and held in place with the rubber band. A large hose clamp also works well for holding the vinyl on the tape roll.

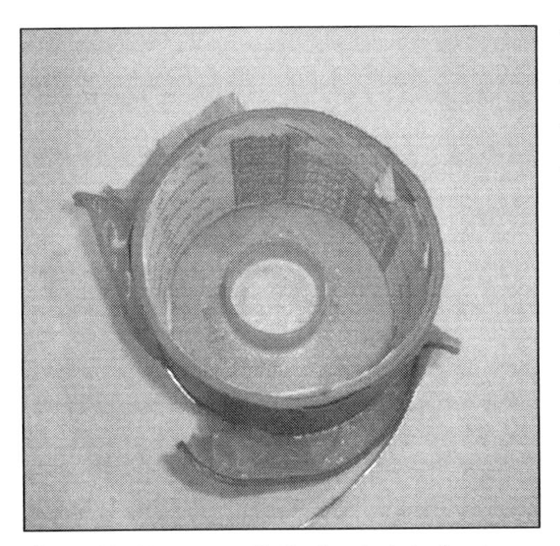

**Figure 23 - Vacuum pulls the heated vinyl onto the form. Turn off the vacuum after the vinyl has cooled enough to maintain its new shape.**

The **form** can be made from a variety of materials. It only needs to be able to stand up to the mild heat and pressure of the thermoforming process. I experimented with several different forms that worked well and finally settled on a very simple approach. The form I use now is a section cut from the top of 35mm film can.

This creates a diaphragm that is very easy to install on your engine after it is formed. Take the lid off a plastic film can and mark around the outside 3/4" down from the top. Cut carefully along the line. The resulting tube will be the form that is used to vacuum-form the diaphragm.

**Note:** Don't start any thermoforming just yet. Refer back to these instructions when you are ready to make your diaphragm.

## Thermoforming Process

We will cut up a vinyl glove and use that as our thermoforming material. The vinyl will be held flat on a small hoop-shaped frame and heated. When the vinyl is warm enough to be shaped it will be pressed over the form and vacuum will be applied. After it has cooled it will hold the new shape.

I will provide you with detailed instructions for thermoforming when we are at that step in the construction process.

# Chapter 8: Tools Needed for These Projects

### *Keeping it simple*

The premise of this book is that you can build a heat of the hand LTD Stirling engine without the aid of a machine shop, and without a lot of money. The tools listed here are the primary tools I keep handy in my tool box when assembling an engine. The process is more difficult than assembling a kit because it requires that you be able to manufacture your own parts from readily available materials.

Before you head to the store and start buying a bunch of tools, understand that there may be ways you can work around some of the more specialized items. For instance, you might be able to stop by a machine shop or an HVAC shop and have your aluminum parts cut for a small price, or even for free. This would eliminate the need for a nibbler in most of these projects.

### *Recommended Tools*

**Nibbler:** (For cutting sheet aluminum to make the pressure chamber sides.) This is a tool designed to cut sheet metal without distorting it. Each of these engine designs requires two pieces of aluminum that are 6" by 6" square. Tin Snips have a tendency to bend and distort the metal as it is being cut, especially if the length of the cut is longer than the cutters on your snips. If you have a good set of tin snips and you can cut metal without bending it, then you can skip this tool. A nibbler can be purchased at better hardware stores for about $28. I got mine on sale at Harbor Freight for $5.

> Possible workarounds: Metal cutting band saw, hacksaw, or a jig saw with a metal cutting blade. You might also be able to have your metal cut to size at a shop that works or sells sheet metal products.

I used the nibbler frequently when I was making engines that required round sheets of aluminum for the pressure chamber parts. I used it very little with these three designs because I purchased material that only required one straight cut to make both pressure chamber side panels.

**Long Nose Pliers:** (aka: Needle Nose Pliers) One or two good pairs of long nose pliers are indispensable for shaping small wire parts. If you have the ability to avoid the cheap tools, you should. The dollar store tools look just like the more expensive varieties, but they will bend when you try to shape high carbon music wire that is used in these projects.

**Diagonal Wire Cutting Pliers:** These are perfect for cutting soft wire parts, however for heavy wire cutting I like to use a pair of electrician's pliers, and I use a file for cutting music wire.

**Electrician's Pliers:** These are a great and versatile tool. They have long heavy handles and can cut through very tough material. They have large flat jaws that are good for working with sheet aluminum that needs flattening or straightening.

**Sandpaper:** I keep a sheet of 80 grit, 120 grit, and 220 grit within close reach while making all of my engine parts. Lay a full sheet on a flat surface and stroke small parts across the surface to shape and polish. This is another area where I recommend that you don't skimp on quality. Good sandpaper looks expensive in the store, but it works so much better, and lasts so much longer than the cheap stuff that it is worth the extra expense.

If you will be working with acrylic (and you want to make it look extra nice) you should also have a variety of *fine* to *very fine* wet-dry emery paper in your collection.

**Electric Drill and a Variety of Bits:** If you are extremely careful you can drill all the required holes free hand. A drill guide or a drill press will greatly improve your accuracy and is recommended, but not required for these projects. A center punch is helpful for locating holes in aluminum, but is not recommended for use on acrylic sheet.

Drilling holes in acrylic will require the use of some modified or specialized drill bits. Regular wood bits cut at a sloped angle and have a tendency to be pulled into the material like a wood screw. This splits and cracks the acrylic. Drilling acrylic successfully requires the use of a bit that has a scraping action, rather than a cutting action. You can modify a standard wood bit to work well in acrylic, or you can buy specialized bits. Bits for masonry or glass use a scraping action to cut, and can be used in acrylic if they are sharp.

**Digital Caliper:** This is a tool that is very useful, but not required. We are building an engine that depends on close tolerances and precise measurements. Having a precision measuring tool helps in this regard. I picked up one on sale thinking it would be a novelty tool, but now I find I am constantly reaching for it. I am very sure I could build a Stirling engine without one, but sometimes you need an excuse to buy a new tool!

**Utility Knife and Blades:** A utility knife or an X-Acto® knife is helpful for cutting and trimming parts. I prefer the utility knife because of the stronger blade. Buy a pack of extra blades and change them often so that you are always working with a sharp knife.

**Straight Edge:** A 12 inch metal ruler will do the trick.

**Square:** It doesn't need to be big. It can even be a small block of wood as long as it is perfectly square.

**Ruler or Tape Measure:** Most of the measurements in these plans are in fractional inches. You can use the scale of your choice. If you use something other than fractional inches you will have to do a little math to make the necessary conversions.

**Coping Saw:** A coping saw with a fine tooth sharp blade is a good way to cut the foam displacer by hand. But you can also use a utility knife, X-Acto® knife, a jeweler's saw, or a jig saw. This tool is optional.

**Cotton Swabs:** Q-Tip or equivalent. These are great for swabbing up glue that oozes out in the wrong place. Also, you can rip the tip off and they make great stirring sticks for mixing epoxy.

**Fine Tip Permanent Marker:** Sharpie or similar type. A fine tipped Sharpie is handy when you need to mark directly on glossy surfaces that don't take pencil or pen. But be careful because these marks are permanent.

**Compass:** This is the type for drawing circles with a pencil, not the gadget that points north! A simple dime-store compass will do the trick. It needs to open far enough to draw the outline of your flywheel.

**Hammer:** (For fine tuning your engine!) A standard duty claw hammer seems to be a handy thing to keep around. You may not need this, but it will be helpful if you need to flatten your sheet aluminum after cutting.

**Adhesives:** I use three types of adhesive: Silicone Glue, Epoxy, and Super Glue®.

The **Silicone Glue** is the same substance as Silicone II caulking compound except it is sold in a small handy tube that can be cut to provide a nice small bead of material. It is more expensive when packaged this way, but very handy. If you have made a pop can engine you probably used high temperature red silicone gasket compound. That is about the best silicone adhesive I have used, but the bright color is not pretty at all.

There are lots of choices when it comes to **Epoxy**. Drying times range from 5 minutes to 24 hours. (I have even seen epoxy recently that says it sets in 60 seconds.) I have been told that glues with longer curing times are generally stronger than the quick setting varieties. I have used both successfully. My opinion seems to change occasionally when it comes to picking my favorite variety of epoxy. There are some jobs where you really want more than 5 minutes to work with the glue joint. For those I will use a 30 minute or 60 minute epoxy. But for most of my work with these engines I really prefer the 5 minute variety that dries clear, or nearly clear. It lets you get a lot more work done in a day because you can glue some parts, wait 15 to 30 minutes, then start to use them. 5 minute epoxy really helps speed up your build time. If I can, I try to buy the epoxy that is sold in two separate bottles or two separate tubes. I find I have a lot of trouble with the epoxy that comes in the double syringe. It is difficult to get equal amounts of each part. And there is some cross contamination in the cap that causes the glue to go bad before I have used it all.

There are several kinds of **Super Glue®** also. The two basic divisions are gel and liquid. The liquid glue requires that your joint fit perfectly with no gaps. It will not fill a gap, and the glue will fail if your joint is not a perfect fit. Gel glues are designed to fill small gaps and are much more forgiving. If you can, buy your super glue at a model shop rather than at the dime store. They will have several varieties and will be able to offer advice on which glue is best for your application. The model shop will also sell a super glue accelerator that will speed up the curing process. This is especially helpful in dry climates.

Here is a good general rule for how to use adhesives in these projects: If you think you might need to take it apart to fix it, use silicone. If you are sure it can stay together, use epoxy. If you use silicone or epoxy on acrylic the surface should be roughed up with sandpaper or the glue will not hold for long. I make as few silicone joints as possible because they do seem to come apart fairly easy.

**Acrylic Cement:** While referred to as cement, it is actually a solvent. It is as thin as water. When applied to an acrylic joint it dissolves some of the acrylic material and then evaporates, leaving a transparent, welded seam.

**Acrylic Tools:** Acrylic can be worked with many of the same tools used for most woodworking projects. These projects will require that acrylic sheet be cut, formed, and drilled. You may be able to accomplish these tasks with tools you already have. See the chapter on Working with Acrylic for a detailed explanation. You must be prepared to safely cut and drill holes in acrylic sheet material that is 1/4" thick.

**Heat Gun:** Some of the designs will ask for small acrylic parts to be heated and bent. A heat gun works very well for this, but is not the only method available. Any heat source that can reach 300° to 350° Fahrenheit will work. A heat gun is also essential if you choose to make your own thermoformed diaphragms. A very hot hair dryer will sometimes meet the need.

**Masking Tape:** Tape will be used to hold glue joints while the glue sets, and to hold temporary assemblies together while parts are being shaped and fine tuned.

**File:** The best way to cut music wire is to score it with a file and then bend it and break it. Music wire is so hard that it can ruin a good pair of wire cutters or pliers.

**Allen Wrench:** There are several shaft collars on each engine. They are operated by a tiny Allen head set screw. If you put a piece of colored tape on the handle of the wrench (like a small flag) it will help prevent you from losing the tiny wrench on your busy workbench.

**Infrared Thermometer**: I purchased a cheap infrared "non-contact thermometer" from Harbor Freight for less than $20. It has proven to be a very valuable tool for measuring the surface temperatures of my engines when they are running. You don't have to have one of these to complete these projects, but it is helpful in measuring your effectiveness.

**Carpenter's Sliding T-Bevel:** A carpenter's bevel is an adjustable gauge for setting and transferring angles. The handle is usually made of wood or plastic and is connected to a metal blade with a thumbscrew or wing nut. The blade pivots and can be locked at any angle by loosening or tightening the thumbscrew. It is a useful tool for building and mounting the pressure chamber, since it has some angles in it that are not square.

**Soldering Tools:** Building the crankshaft for Engine #3 calls for a sweat soldering technique that uses a propane torch. You can substitute a good epoxy for this step if you do not have access to soldering tools. The epoxy will not be as strong as a soldered connection.

# Chapter 9: Engine #1 - The Reciprocating Stirling Engine

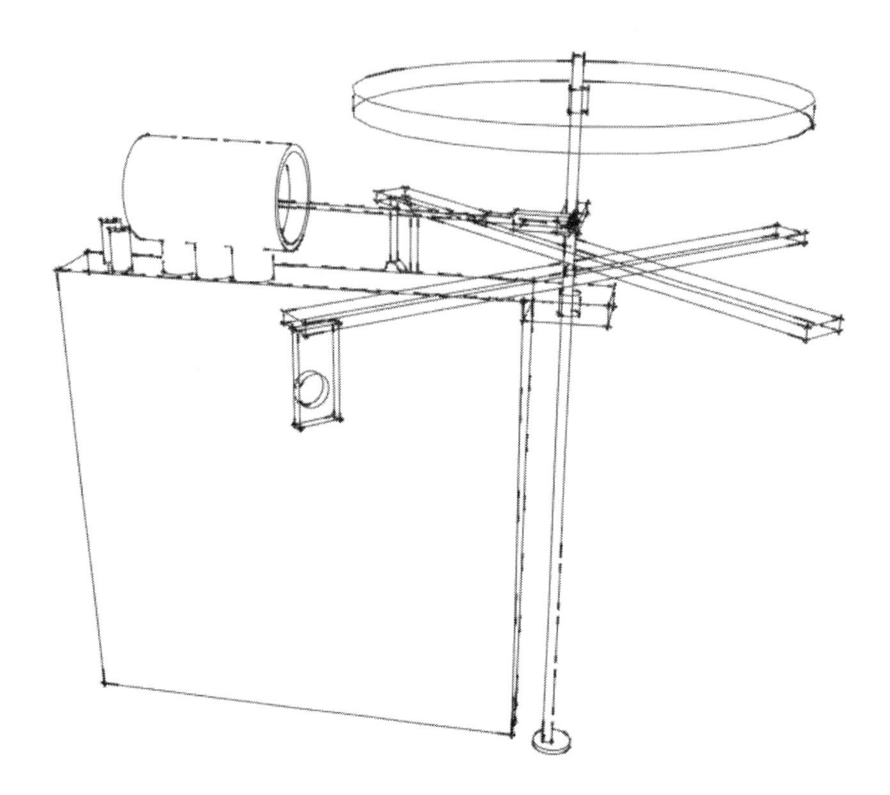

I have sometimes called this, "The self adjusting magnetic drive reciprocating Stirling engine." It is not a very catchy name, but it is very descriptive. The flywheel on this engine does not rotate all the way around. It will rotate about 90° before stopping and then moving in the opposite direction. The motion and the sound it makes often reminds people of a ticking clock.

When I say it is "self adjusting", I do not mean to imply that it will tune itself to optimal running conditions. You still have to do that. But what this engine can do (once it is tuned) is make minor adjustment for changing pressure differentials. Most Stirling engines will compensate for these changes by running faster or slower. This engine also compensates by being able to adjust the length of the power stroke to use more energy when it becomes available.

This is a true "heat of the hand" engine. I have successfully operated this engine from just the heat of my hand on many occasions.

## Parts List

The table below represents the retail prices in March of 2010. Many of the supplies I purchased came in large quantities that will provide enough material to make several engines. Because of that, I have listed 2 prices. The first column is the price I paid for the material. The second column represents the adjusted cost for the amount of materials that were actually used in this engine.

| Engine #1 | | | |
|---|---|---|---|
| **Item** | **Count** | **Retail** | **Actual** |
| 6" x 12" x 0.064 Aluminum Sheet | 1 | $ 7.29 | $ 7.29 |
| 1/8" (0.125) Music Wire | 1 | $ 1.16 | $ 0.29 |
| 1/16" (0.062) Music Wire | 1 | $ 1.88 | $ 0.31 |
| 3/32" (0.093) Clear Acrylic | 1 | $ 2.79 | $ 0.20 |
| 1/4" (0.220) Clear Acrylic | 1 | $ 14.97 | $ 3.74 |
| 1/4" Brass Barb Splicer | 1 | $ 2.18 | $ 2.18 |
| Vinyl Gloves (10 pk) | 1 | $ 3.97 | $ 0.40 |
| 1/8" Steel Shaft Collar | 4 | $ 1.00 | $ 4.00 |
| 1/4" ID Vinyl Tubing (10') | 1 | $ 3.11 | $ 0.05 |
| 1/2" OD Vinyl Tubing (10') | 1 | $ 4.15 | $ 0.03 |
| 35mm Film Can | 2 | $ - | $ - |
| Neodymium Magnets 1/16" x 1/2" | 4 | $ 0.45 | $ 1.80 |
| JB Weld Epoxy | 1 | $ 4.97 | $ 0.99 |
| 5 Minute Epoxy | 1 | $ 3.97 | $ 0.79 |
| Clear Silicone | 1 | $ 3.89 | $ 0.97 |
| 3/16" Foam Poster Board | 1 | $ 3.62 | $ 0.08 |
| Black Spray Paint | 1 | $ 0.97 | $ 0.24 |
| Helium Filled Balloon | 1 | $ - | $ - |
| Penny | 1 | $ 0.01 | $ 0.01 |
| 4-40 x 1 Machine Screws and Nuts (10pk) | 1 | $ 0.99 | $ 0.33 |
| Acrylic Cement | 1 | $ 5.30 | $ 0.53 |
| 1/8" ID Glass Bead | 1 | $ 0.10 | $ 0.10 |
| | | $ 66.77 | $ 24.35 |

Since I am a bargain hunter, I will always shop around for the best prices. If you are lucky enough to find scrap acrylic for $1.00 a pound and a good discount hardware store you should be able to bring that $66.77 figure down to about $42.00.

## Finding the Parts

What follows here is a brief description of how I chased down the parts for this engine. I almost always chose the least expensive option when I had the choice.

The **Aluminum Sheet** was purchased at a local hobby shop. This particular store sells radio controlled airplanes, cars, helicopters, and things of that nature. They also sell metal parts for people who make their own models. The brand name on the metal plate is K & S Engineering. There were several shelves of K & S parts in the store. They sell tubing, shaft, and sheet metal in aluminum, brass, and copper. I have also seen K & S Engineering display racks at one of our local True Value hardware stores. I was able to purchase a 0.064 x 6" x 12" sheet of aluminum for $7.29. After building this engine I stumbled across enough aluminum to make 6 to 8 engines at the Boeing Surplus Store for $8. They were selling scrap aluminum for $1.50 per pound. Unfortunately that store is now closed.

The **1/8" Music Wire** and **1/16" Music Wire** came from the same hobby shop. It is sold in 3 foot lengths. This project only uses about 8 inches for the main drive axle, so there will be plenty left over for your next project. Music wire is also called "piano wire" or "pushrod wire" in the model airplane world. It is an excellent choice these projects because it is very stiff. It has a high carbon content, it is accurately sized and has a polished surface. I have also seen 1/8" rod for sale at the home stores, but it is usually not music wire. The product I saw there was soft metal and it lacked the polished finish. I highly recommend the music wire for these applications.

I have purchased **Clear Acrylic** sheets from two different stores. The prices quoted above are from a recent trip through Home Depot, which is where I purchased my first acrylic for these projects. I discovered I could get a much better selection and far better prices at a store that specializes in plastics. Tap Plastics in Seattle sells their scraps for a dollar a pound. I walked out of the store with a sizeable stack of several thicknesses ranging from 3/16" thick to 1/2" thick. The whole stack of scrap cost about the same as the one sheet of 3/32" material from the home store. One of these Stirling engines will require a sheet of 1/4" acrylic measuring about 10" x 12", and about 2" x 10" of 3/32" acrylic sheet. The 1/4" material at Home Depot measures 18" x 24", which is about enough to make 4 of these engines.

As I mentioned earlier, "Plexiglas" is one of many brand names for clear acrylic sheet plastic. When it comes to sheets of clear plastic there are two popular materials: acrylic and polycarbonate. Polycarbonate sounds like a good choice at first. It is stronger, harder, and has better scratch resistance. Unbreakable eyeglass lenses are made from a form of polycarbonate. Acrylic has some qualities that make it a better choice for these projects. Acrylic is cheaper, and acrylic is easier to bend and form than polycarbonate. When acrylic is heated it becomes soft and flexible. It has the consistency of wet leather. It can be bent, shaped, and molded into a variety of shapes. Hold it in position until it cools and it will retain the new shape.

You may not have access to an inexpensive outlet like Tap Plastics, so I have listed the prices of the more common supplier in the parts list. Tap Plastics does sell many of their products online and they are a good source for Acrylic Cement if you can't find it locally.

Figure 24 - Cut a brass elbow to make two vent nipples.

The **1/4" Brass Barb Splicer** is a fitting made for splicing vinyl tubing. One fitting can be cut to create two vent nipples. I went to my favorite hardware store and told them I needed some 1/4" tubing and fittings. I was directed to a wall of small fittings of various sizes and shapes. These fittings are available in plastic or brass. I initially purchased plastic fittings because they were cheaper, but then I decided to splurge for the brass because I thought it would look nice. The engine needs two 1/4" fittings attached to the pressure chamber. These serve two purposes. First, they allow you to equalize the pressure in the pressure chamber to compensate for barometric pressure changes. Second, with two vents it is possible to easily fill the chamber with helium. Filling the engine with helium provides a dramatic increase in engine performance, especially at low temperature differentials.

I chose to buy a 90 degree elbow that was made to accept 1/4" ID plastic tubing. I cut the elbow in half to create two nipples that will be used for vents. A piece of vinyl tubing will be used to seal the vents while the engine is in use.

The **Vinyl Gloves** came from the paint department at Home Depot. There are lightweight vinyl gloves and there are heavyweight vinyl gloves. Get the heavyweight gloves. The thermoform process will make them thinner, and the heavy material will last longer.

Be careful when shopping for gloves that you do not buy gloves made of nitrile or latex. Nitrile and latex will not work for the thermoforming process that will be used to shape the diaphragm needed to drive the flywheel.

The **1/8" Steel Shaft Collar** is a key component to the simplicity of this design. It fits perfectly on the 1/8" rod that is used for the axle. Anything that needs to be mounted to the axle is first glued to the steel collar and then fastened to the axle with the setscrew in the collar. It just takes a few twists with an Allen wrench and any of these parts can be easily moved for adjustment. I found my steel collars (also called "shaft collars") at a commercial hardware store called Tacoma Screw Products. Shaft collars are also available online, or at any well supplied hardware center.

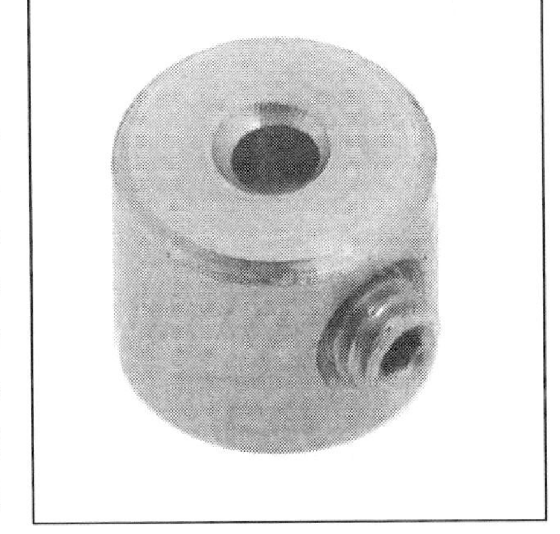

Figure 25 - Steel Shaft Collar

**1/4" and 1/2" Vinyl Tubing** is sold by the foot at some hardware stores. If you can find it being sold that way you will save a bit of money. Home Depot sells it in 10' rolls but you will only be using a few inches of each. You will need about 2"of 1/4" tubing and about 2" of 1/2" tubing. The smaller tube is used to seal the brass vent nipples on the pressure chamber. It also aids in filling the chamber with helium. The larger diameter tubing is used to connect the pressure chamber to the film canister that holds the diaphragm drive. Buy at least a foot or two of each size. It is really pretty affordable stuff and comes in handy.

The **35mm Film Can** was obtained for free. I went to a one-hour photo processor and asked if they had any extra film cans. The nice man behind the counter gave me 8 of them! There are two popular styles. One is black with a grey press-on lid, and the other is frosty white with a press on lid. Some film cans have a lid that fits inside of the can while others have a lid that is larger and grasps the outside of the can. Any of those configurations will work.

I ordered **Neodymium Magnets** online from K & J Magnetics (http://www.kjmagnetics.com/). Neodymium magnets (sometimes called "rare earth magnets") are incredibly powerful for their size. In fact, they come with a safety warning. The magnets I purchased are 1/16" x 1/2" round magnets grade N42. I have also used 1/8" x 1/2" round magnets with great results. The larger magnets develop over 16 pounds of pull, which makes them incredibly difficult to take apart at times. The smaller magnets are cheaper and easier to work with. I have not found a reliable local source for specialty magnets. K & J Magnetics delivered my order promptly and I was happy with their service.

**Figure 26 - 1/16 x 1/2" round magnet.**

I purchased the **Epoxy** from the paint section at Home Depot. I usually keep two different kinds on hand. *JB Weld 60 Minute Epoxy* is known for its ability to bond well to metal and it also withstands high temperatures well. I used JB Weld on my first pop can Stirling engine and it is still running strong years later. I also keep some 5 minute clear epoxy on hand because the short cure time cuts down on time spent waiting for glue to cure. The clear epoxy looks nice with the acrylic parts. Read the label carefully on the various brands. I purchased one brand of 5-minute epoxy that said it reached its full strength in 24 hours, and another brand that reached its full strength in just 1 hour. Not all 5-minute epoxies are the same. As I mentioned once before, I strongly recommend you always buy the epoxy that comes in two separate tubes. The conjoined twin syringes look good in the store but they cause a lot of problems.

The **Clear Silicone** came from the same big home store. You can purchase this in small convenient tubes, or in large cylinders that fit in your calking gun. They are both about the same price, but the smaller tube is much easier to work within the tight confines of these projects. Silicone II is the better choice. You will see some labeled for kitchens, some for windows and doors. It doesn't really matter which you get because they all contain the same thing. The clear stuff will look much nicer on your project.

Hint: Silicone does not stick to the shiny surface of acrylic sheet very well. Carefully rough up the surface with some fine sandpaper and it will adhere a little better.

My wife enjoys art and has a studio of her own. I could tell you that I acquired the **3/16" Poster Board** for free because I found it in her studio. But it had a $2.99 price tag on it from the arts and crafts store. I have also seen this offered for sale in dollar stores, but the quality suffers a bit at that price. I like the paper coating on the poster board. It makes it a little heavier, but also makes it quite a bit more durable. The poster board is sometimes labeled 1/4". If you see it labeled that way, measure it before you buy it. If it is a full 1/4" thick it will be necessary to modify the dimensions of the pressure chamber to accommodate it.

The **Black Paint** came from the big home discount store. Painting improves the performance of your Stirling engines. You will want to paint both sides of both of the aluminum plates black. Those who like to do things right will add a coat of primer paint that is made especially for aluminum. I have built these with and without primer, and they work fine both ways.

I picked up the **Helium Balloon** (filled) on my way out of the Red Robin® restaurant after dinner. They usually offer the balloons to the children. They didn't seem to mind me getting one. You should at least be kind enough to buy a burger while you are there. You enjoy a great meal and get a supercharged engine at the same time. The Applebee's® restaurant in our neighborhood also gives away balloons to their customers. If you don't have access to a free source of helium, I suggest picking up some filled balloons at the dollar store. Don't pick up your balloon until you have finished your engine and you have it running. The balloons don't last very long. The helium molecules are small enough that they can leak out through the latex balloon material.

Do I really have to tell you where to get a **Penny**? If you don't have one in your pocket, maybe you can borrow one from a friend. The penny is used as part of the bearing. The vertical axle will rest on the penny.

The **Machine Screws** also came from Home Depot. They are sold in a small package that includes 10 nuts and 10 bolts.

The **Acrylic Cement** was purchased at Tap Plastics in Seattle. If you do not have access to a specialty plastics store near you, I recommend you look them up online at http://www.tapplastics.com/.

The **Glass Beads** were purchased at Shipwreck Beads near Olympia, Washington. I have seen suitable beads at other craft stores as well. Shipwreck Beads sells their products over the Internet, but their inventory is so vast it might be difficult to find exactly what you are looking for. I took a section of the axle material into the store and experimented with different beads until I found some that were a good fit. My wife really enjoyed hanging out and shopping while I was on my geeky bearing hunt.

## Conceptual Drawings

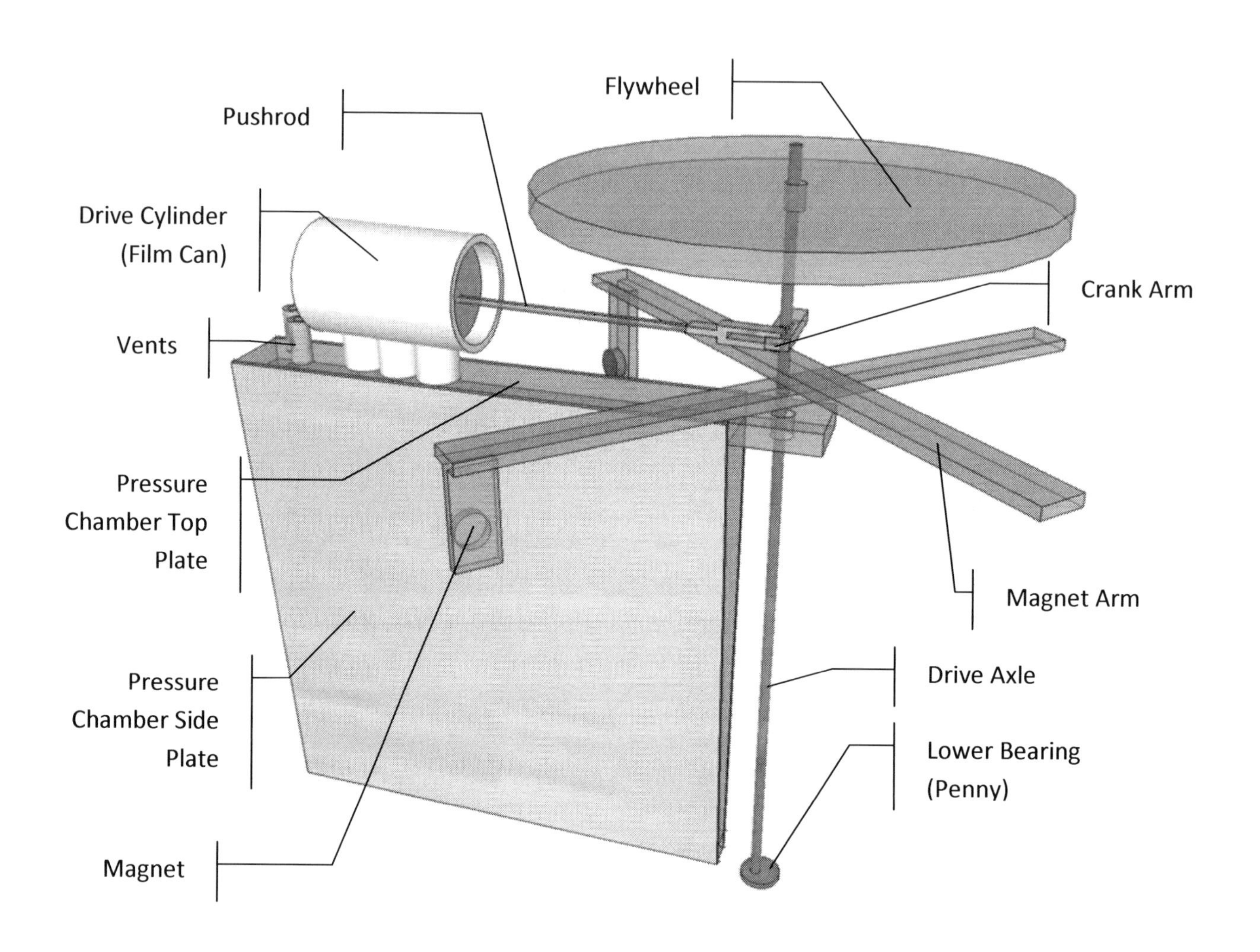


**Figure 27 - These are the main components of Engine #1.**

### *Pressure Chamber Dimensions*

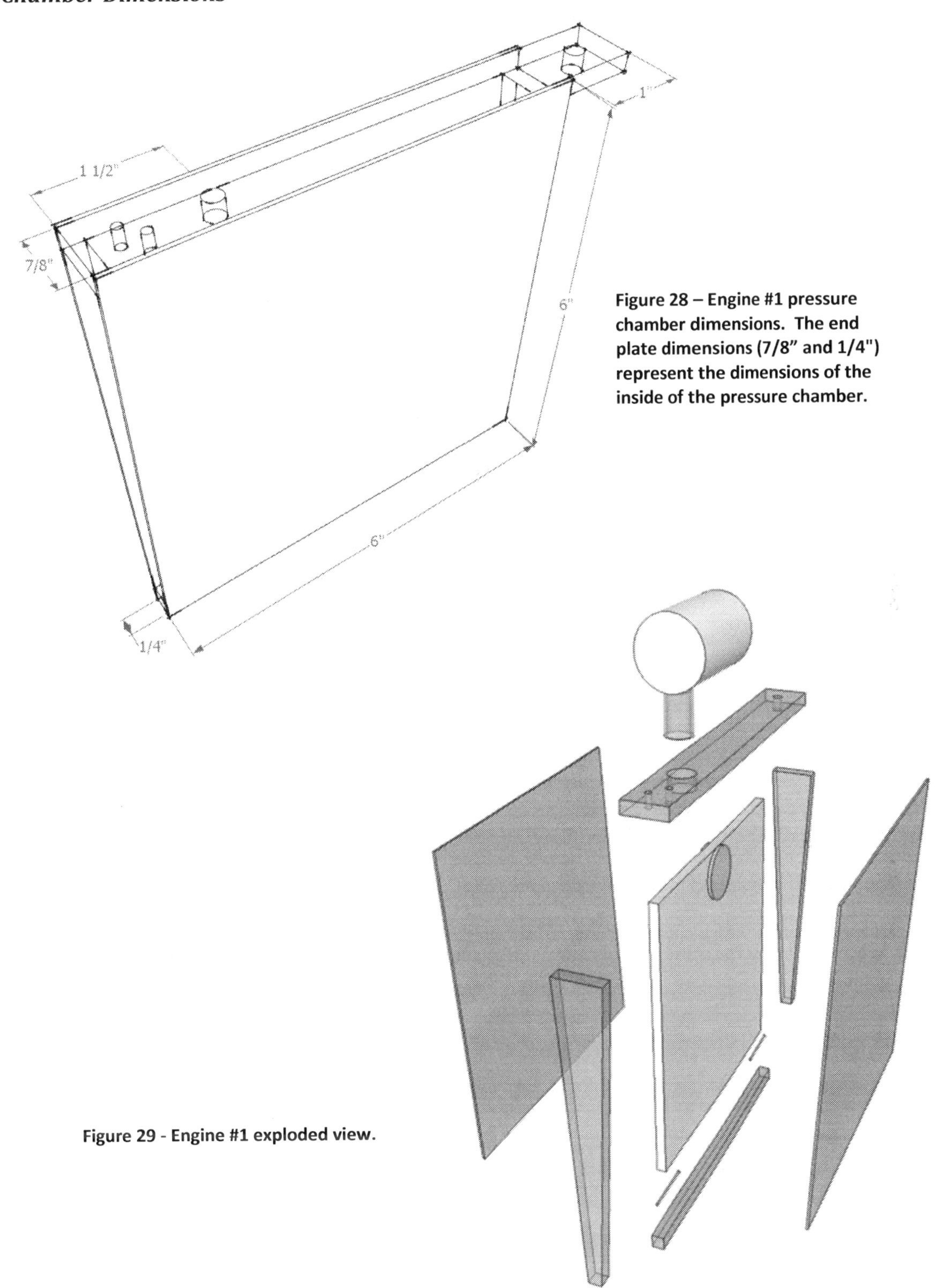


**Figure 28 – Engine #1 pressure chamber dimensions. The end plate dimensions (7/8" and 1/4") represent the dimensions of the inside of the pressure chamber.**

**Figure 29 - Engine #1 exploded view.**

### *Acrylic Part Template Dimensions for Engine #1*

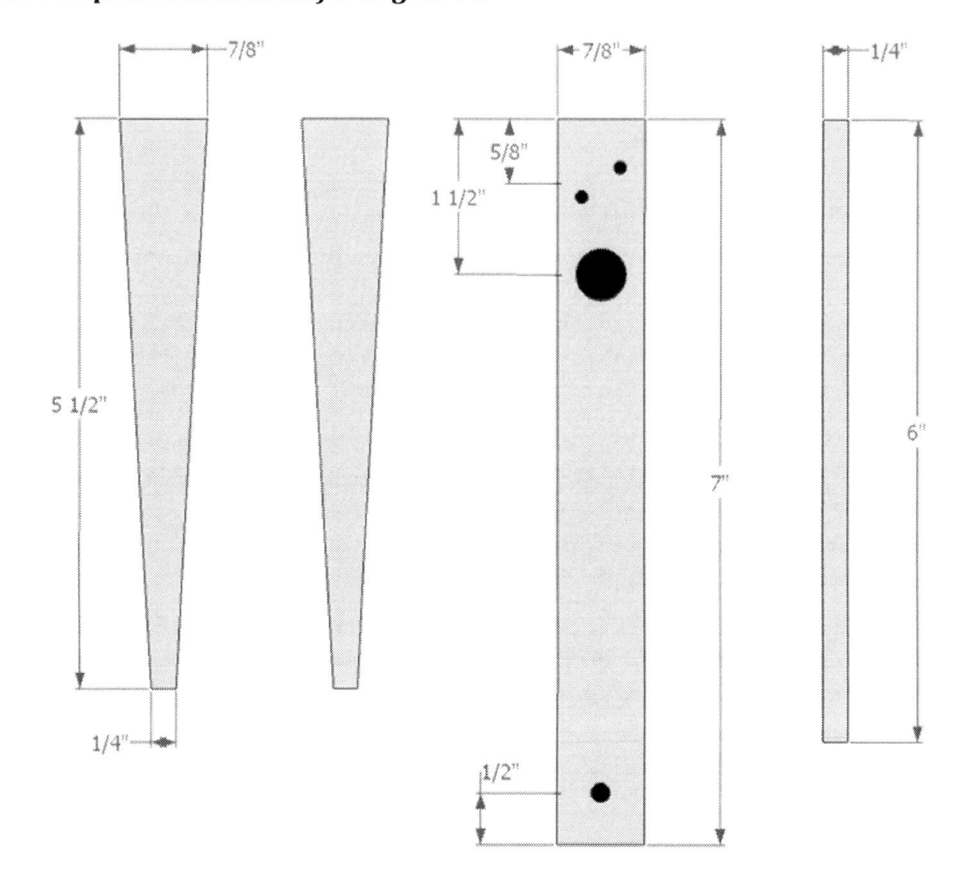


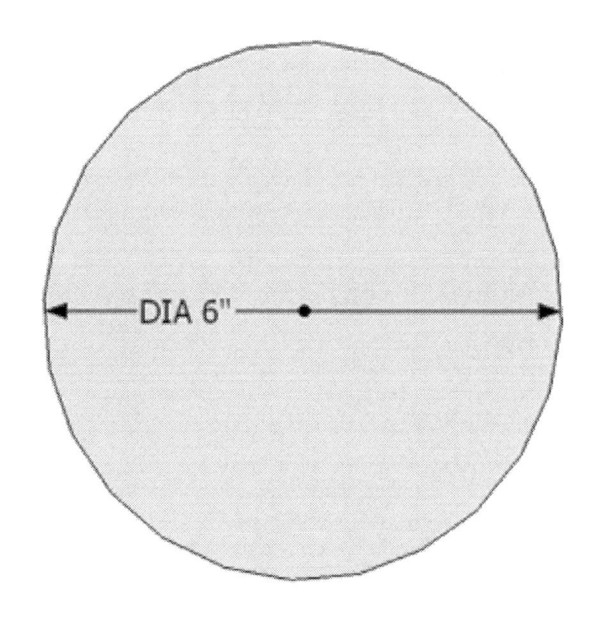


**Figure 30 - Engine #1 acrylic part templates. The hole for the axle in the rectangular pressure chamber top plate is 3/16". The vent holes are 1/8". The hole for the drive cylinder is 1/2" and is intended to be a snug fit for the tubing used to connect the drive cylinder.**

**The flywheel is a 6" diameter disk with a 1/8" hole in the center.**

## Step by Step Instructions

### *Draw Plans and Create Templates*

It is time now to take out paper, pencil, and a ruler and to draw your own set of full sized plans. Use the dimensions listed in these illustrations to create a set of full sized drawings. These will be cut out and used as patterns for creating the acrylic parts for the pressure chamber and the flywheel. (All of the illustrations in this book are drawn to scale, but they are not full size.)

The flywheel will be cut from the same material as the pressure chamber parts. You should keep that in mind as you contemplate how you will lay out your patterns on your material for cutting.

**Figure 31 - The templates for two pressure chambers are laid out on the acrylic sheet in preparation for cutting. Making full sized templates is very useful in this process.**

**And here is another tip:** You will notice as you compare the first two designs that the pressure chambers are almost identical. Engine #1 has a single tab protruding from one side at the top to hold the drive axle. Engine #2 has a tab at both ends to hold two axles. The top acrylic piece is the only difference between those two pressure chambers. If you build this engine with the extra tab on the top of the pressure chamber you can use it to assemble either engine design. I prefer having both designs so I can compare them side by side. But you can reconfigure your engine to run in either mode in just a few minutes if you have the second tab on your pressure chamber.

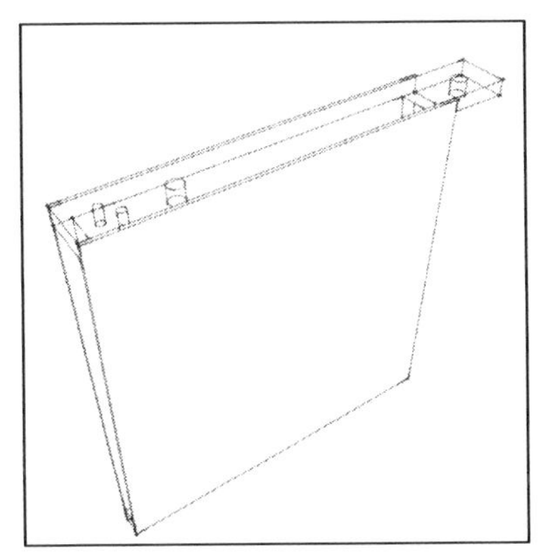

**Figure 32 - The pressure chamber for engine #1 has a tab extending from one end that holds the driveshaft bearing.**

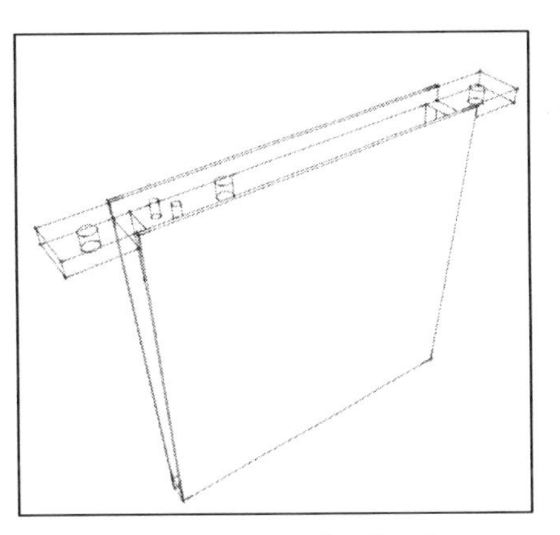

**Figure 33 - The pressure chamber for engine #2 has the same dimensions as the first one except it has a tab on both sides for holding bearings.**

### *Cut and Paint the Aluminum for the Pressure Chamber Side Walls*

You will need to cut two 6" squares of aluminum sheet. The aluminum I purchased for this project comes in a sheet that measures 6" by 12". Only one simple cut was required.

**Figure 34 - Pressure chamber side walls are painted black on both sides except for the inside border (shown here).**

The aluminum must be of a heavy enough gauge so that it will not easily flex. Do not use thin flimsy metal as it will not perform as well. The material here has a thickness of 0.064". Aluminum that has a thickness near 0.04" will also work. If it is much thinner than that it becomes too flexible for a pressure chamber.

I don't recommend the use of tin snips for cutting flat sheets of metal. The tin snips will bend the metal on one side of the cut.

You can cut this by hand with a hack saw if you are patient, but it will be a challenge because a hack saw is not really intended for cutting material that is 6" wide.

I recommend either cutting it with a nibbler or with a metal cutting blade in a saber saw. If you use the nibbler you will need to flatten the edge of the metal after it has been cut, but that is easy to do. If you use a saber saw you will want to use a board and some clamps to hold the metal securely to your workbench and then use the board as a cutting guide. Always secure metal parts with clamps before attempting a cut with a power tool. You don't want these sharp objects flying through the air at high speed! It can get quite noisy, so hearing protection is recommended.

Please refer to the chapter, "Working with Aluminum" for a detailed explanation of cutting and finishing techniques.

Use masking tape to prevent painting the area around the inside edge of each sheet where the acrylic parts will be attached to the aluminum. This will help provide a strong glue joint. Glue sticks better to bare aluminum than it does to painted aluminum. Paint both sides of the aluminum sheets black.

The black paint really does make a difference in engine performance. I have made Stirling engines with black paint, white paint, and no paint. I now paint all sides of my aluminum sidewalls with flat black spray paint.

### *Cut the Acrylic Parts for the Pressure Chamber Bottom, Ends, and Top*

**Figure 35 - Pressure chamber top, bottom, and end pieces cut from 1/4 inch acrylic sheet.**

The pressure chamber dimensions are made to accommodate the material chosen for the displacer. My displacer material is about 3/16" thick. If you are able to locate material for your displacer that is 3/16" thick then just use the dimensions provided in the drawings at the beginning of the chapter. If for some reason you cannot find an exact match for the displacer material you must adjust the dimensions of your pressure chamber accordingly. i.e., If your displacer is 1/4" thick then you should make the pressure chamber 1/16" wider at both the top and the bottom.

Lay out the full scale pattern for each of the acrylic parts and attach it to the surface of the acrylic for cutting. Do not remove the protective coating from the acrylic sheet until after the parts are cut. Leave enough space between the parts to accommodate the width of the cut that will be caused by your saw. I used a saber saw for cutting the parts you see in these illustrations.

Use a sanding block if necessary to make all cuts in the acrylic flat and straight. Both end pieces (the tapered ones) must be the identical shape. Tape or clamp these pieces together when sanding the edges in order to maintain a consistent shape. Only sand the cut edge. Do not sand the shiny flat surface of the acrylic. If cutting leaves a buildup of melted acrylic you can sometimes scrape it off. Use another piece of acrylic as a scraper to remove the rough particles that may be stuck to the cut edge.

The long edges of the top plate and the bottom plate should be angled to match the angle of the ends. If you don't have the ability to cut this angle reliably, cut them square and then sand them to the correct angle. It is possible to sand the edges after the frame is cemented together by rubbing the completed frame on a sheet of flat sandpaper.

Drill the holes in the pressure chamber top plate prior to assembly. The holes are less likely to cause a crack or a break in the acrylic if you drill them before cutting. The holes for the bearings are 3/16". The vent holes are 1/8". The hole for the drive cylinder is 1/2"

### *Assemble the Pressure Chamber – Leaving One Side Off*

You may begin to assemble the pressure chamber when the paint is cured on the aluminum parts and the acrylic framework pieces have been cut to size.

Figure 36 - Hold the acrylic pieces so that nothing touches the joint as you join the parts.

Dry assemble all parts first to guarantee a good fit. Shape or replace any parts that are not fitting well.

Assemble the acrylic frame using the capillary cementing technique described in chapter 6. This provides an attractive and sturdy joint for the acrylic frame. When the frame is complete and the joints have cured you will attach it to one of the aluminum side plates. (The other aluminum plate will be glued on after the displacer is assembled and placed inside.)

I learned the hard way that you don't want any clamps or braces (or even tape) to be touching the acrylic joint when joining with solvent. The solvent will run under the tape or the clamp and disfigure your beautiful acrylic surface. The picture here shows how I held the parts square while cementing them together so that nothing would be touching the joints when the solvent was applied.

I recommend using a slow cure epoxy like JB Weld for attaching the aluminum side plates to the acrylic framework. The slow cure will provide you with plenty of time to work with the joint so that you won't be rushed by a five minute time limit like you would with faster curing glue. Also, JB Weld is very thick and sticky, which is helpful for this operation. You want to make sure you don't get too much glue oozing out of the joint. Large globs of glue on the inside of the pressure chamber can interfere with the operation of the displacer after the engine is assembled. Carefully remove any excess glue if this happens.

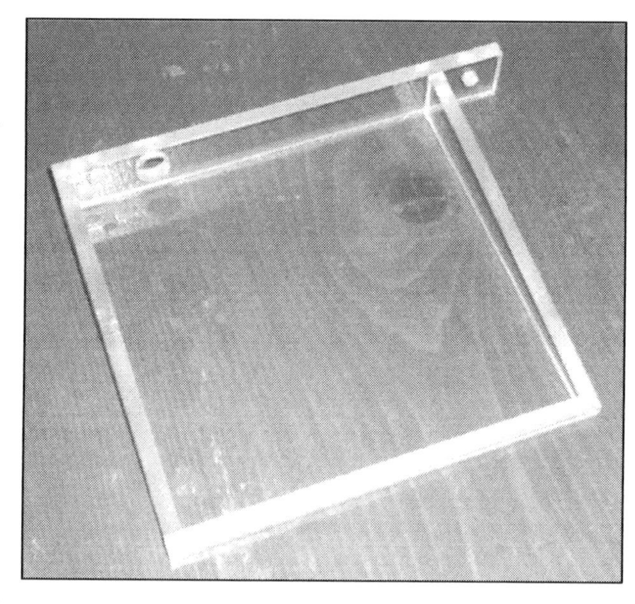

Figure 37 - Pressure chamber frame assembly.

You will not be able to wipe the epoxy off of the glossy surface of the acrylic if you make a mistake. If you dribble a little glue on your acrylic, and it is not in a place that will interfere with the operation of moving parts, my advice is you leave it there. Trying to clean it off will probably only make it look worse.

### *Build the Displacer*

The Displacer will be cut from 3/16" thick poster board. Two thin neodymium magnets will be attached to the center of the top edge, and two small wires will be attached at the bottom. The wires will extend slightly past the end of the displacer and will prevent the displacer from getting close enough to touch the acrylic end plate. The wire also provides a smaller footprint for the displacer and reduces friction that would be caused by having the entire edge of the displacer on the bottom of the pressure chamber.

The dimensions of the displacer will vary slightly depending upon the thickness of the acrylic used to make the end plates. The displacer should be cut so that there is about 1/8" clearance on all sides when it is inside the pressure chamber.

**Figure 38 - Small wires are attached to the bottom edge of the displacer. The tape is removed after the wires are glued in place.**

Measure the inside height and width of the pressure chamber. Cut the poster board to be 1/4" shorter than the inside of the pressure chamber (this will make 1/8" clearance on each side).

Check the fit and mark which side of the displacer will be at the top. Attach a thin (1/16") neodymium magnet to each side of the displacer, centered along the top edge. My magnets were 1/2" round and 1/16" thick. Use silicone or epoxy to glue the magnet to the outside top edge (centered). The magnets must be installed so that they are attracting each other. This is very convenient as it makes it possible to glue them to the displacer without a clamp. The direction of the poles does not matter as long as the magnets are set to attract.

I have also built these with a single magnet imbedded inside the foam core of the displacer material. You can do it this way if you can't find thin magnets. You should understand that the magnets serve a second purpose when mounted to the outside edge of the displacer. They prevent the displacer material from resting flush against the aluminum sidewall. If you choose to place your magnet inside the foam core you will have to add a small spacer (such as a drop of glue) to the top edge of the displacer so it will always stand off from the side wall about 1/16".

Cut two pieces of thin wire to a length of 1". Use sandpaper or a file to remove any burrs from the cutting process. Use tape to temporarily attach one to each bottom corner of the displacer. They should be taped to the end opposite the magnet, so that they rest flat on the bottom of the pressure chamber when assembled.

The ends of the wire should protrude 1/16" on each side. This will act as a "stop" to prevent the displacer from getting too far out of center.

Gently place the displacer assembly inside the pressure chamber to inspect the fit. If both of the short wires are touching the sides of the frame at the same time they are too long. The displacer must be able to rock back and forth with no obstruction. Hold the second side plate in place and check to make sure the displacer clears all sides by about 1/8" as it moves back and forth.

**Figure 39 - Displacer is in position and glue has been applied to the acrylic frame. The magnet is at the top (wide side) and the two wires are at the base (narrow side).**

When you are happy with the fit, use epoxy to glue the wires in place on the bottom edge of the displacer. Use tape to hold the wires in place. Set it aside until the glue is completely cured, and then remove the tape.

### *Attach the Second Pressure Chamber Side Plate*

This is one of the more complicated and tricky parts of this assembly process. The displacer must be placed inside the pressure chamber so that the magnets are at the top (the wide end) and the wires are at the bottom (narrow end). The side plate must be glued in place to make an air tight seal, but still allow the displacer to move freely from side to side.

If glue oozes from the joint and sticks to the displacer, and the displacer can't move freely, the engine will not run.

Here is how I do it:

- Place the partially assembled pressure chamber on a flat surface with the open side facing up.
- Place the displacer inside the pressure chamber.
- Place the second aluminum side plate in place (with no glue yet) and inspect the fit. It should be a flush tight joint.
- Remove the side plate and apply a thin coat of JB Weld 60 minute epoxy to the acrylic frame side of the joint.
- Press the aluminum side plate against the glue and apply enough pressure to squeeze the excess glue from the joint.
- Immediately remove the aluminum side plate and inspect the glue joint to make sure you have an air tight seal. You should see a continuous trail of wet glue all the way around the aluminum side

plate. If there is excess glue oozing into the interior of the pressure chamber remove it now. Be careful you do not accidentally glue the displacer to the inside of the pressure chamber.

- Replace the side plate and gently clamp it with a rubber band or weighted object until the glue is dry.
- Keep the pressure chamber flat until the glue is cured. Do not pick it up or try to make the displacer move while the glue is still sticky. If you cause your displacer to move while there is still tacky glue it could glue your displacer to the inside of your engine, and then your engine will not work.

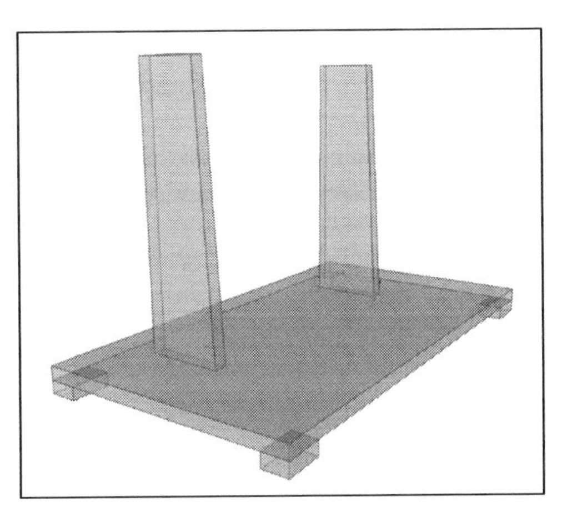

**Figure 40 - The stand holds the pressure chamber upright.**

### *Build a Stand for the Pressure Chamber*

At this point you should have an assembled pressure chamber that has a clear acrylic frame (top, bottom, and ends), two black aluminum side plates, and a displacer. The displacer should move freely from side to side, and you should be able to make it move by holding another magnet near one of the magnets on the displacer. It is now time to build some type of apparatus that will hold it upright and will hold the bottom bearing for the drive axle.

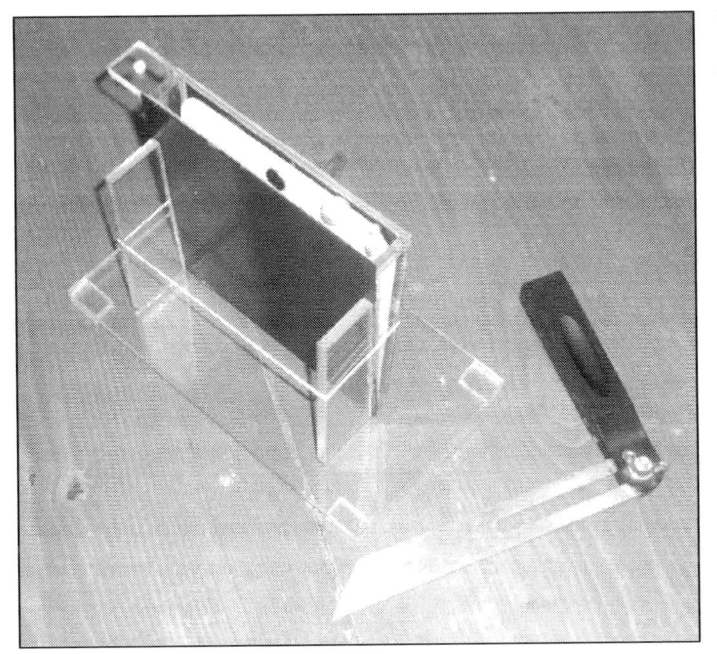

**Figure 41 - Build a stand with angled vertical braces that will hold the pressure chamber with its centerline perfectly vertical.**

The base can be built of wood or acrylic. Acrylic is only slightly harder to work with and makes a very attractive stand. I have used both materials and favor the acrylic for its good looks. The stand needs to provide a wide base for the assembled engine so it does not fall over. It needs to be at least one inch longer than the pressure chamber. For this engine I used a piece of 1/4" acrylic sheet that measures 5" by 9". I cemented four small scraps of acrylic to the bottom of the base to act as feet. These are intended to prevent the bottom of the acrylic base from getting scratched.

The two vertical bars are approximately 1 1/2" by 5". One of the long sides on each vertical brace is cut at an angle so that the pressure chamber will stand with its centerline perfectly straight up and down. The vertical braces must be set as close to the edge of the pressure chamber as possible. Set them so that the outside edge of the brace is flush with the end of the pressure chamber.

You can find the angle for the vertical brace by tracing the same template you used to cut the pressure chamber end plates. You can use a carpenter's bevel tool to check your angles prior to final assembly. If the centerline of the pressure chamber is vertical, the angle will be the same on both sides.

### *Attach the Pressure Chamber to the Base*

The cool side of the pressure chamber will rest against the vertical braces of the stand. Later when you assemble the drive mechanism the pushrod will be attached on the warm side of the engine.

Use clear silicone to attach the pressure chamber to the base. Apply a small bead of silicone adhesive along the contact area of the base and the pressure chamber. Check the angles on both sides of the pressure chamber and when they are correct, use tape or a rubber band to hold the pressure chamber in place until the silicone has set.

Silicone is used because it is not necessarily permanent. If you ever need to disassemble this joint to correct the angle you will be glad you used silicone.

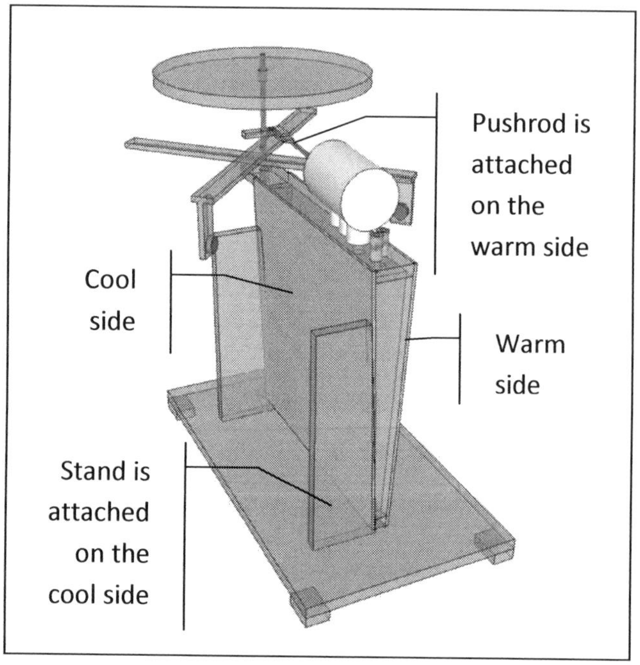

Pushrod is attached on the warm side

Cool side

Warm side

Stand is attached on the cool side

**Figure 42 - The stand is on the cool side of the engine. The pushrod is attached to the warm side of the engine.**

**Note:** The centerline of the pressure chamber must be perfectly vertical! Both sides of the pressure chamber must slope at the same angle. This is critical to the balance of the engine. If these angles do not match, the displacer will not move easily in both directions. Have I said this before?

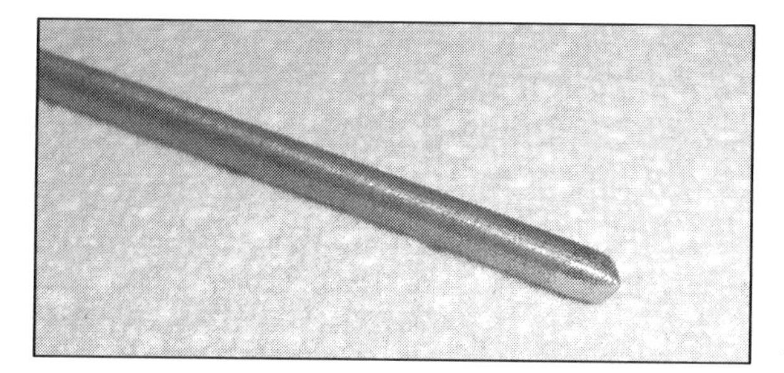

**Figure 43 - The bottom end of the axle is sharpened to a dull point to reduce friction.**

### *Cut and Shape the Drive Axle*

Cut a 9" length of 1/8" music wire. The top end should be finished smooth with a file or a sanding block so that all the burrs are removed. Use a file or sanding block to sharpen the other end of the axle so that it resembles a dull pencil. The goal is to make a small smooth tip on which the drive assembly will ride. A small smooth point will reduce the friction at the spot where the axle rides on the copper penny. If the tip is too sharp it will bore a hole into the penny and increase friction. If you don't sharpen the tip of the axle you will experience more friction as a result of the increased contact area between the end of the axle and the coin.

### *Create the Drive Axle Bearings*

Figure 44 - Heat the acrylic with a heat gun and bend it into shape with scraps of wood.

The lower bearing consists of two pieces: A small bracket to prevent the axle from getting out of alignment, and a penny on which it rides.

Prepare the penny by putting a thin coat of silicone on the back side of the penny and let it cure. After it cures, the silicone will act as a non-skid surface and prevent the coin from wandering while the engine is running.

The bracket is not intended to touch the drive axle. It is only there to prevent the axle from moving out of alignment, which would damage the top bearing. The bracket seen in the photos was bent from a small piece of 1/4" acrylic, and the hole was drilled using a carbide bit. The bracket has to allow the penny to be placed under the end of the axle. The axle will ride on one of the flat smooth areas surrounding Lincoln's head.

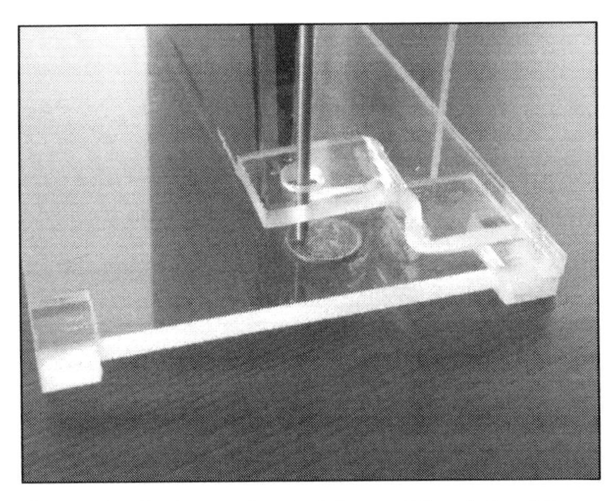

Figure 45 - The keeper bracket does not touch the drive axle, but acts as a safety to prevent misalignment. The drive axle rests on a penny.

I used a heat gun to warm the bracket, and pressed it into shape with some scraps of wood. It can be glued to the base by a capillary cementing process or with epoxy. If you use epoxy to glue acrylic you must rough up the surfaces with sandpaper first. If you don't do that the glue will not hold.

The top bearing is a glass bead. The hole in the bead is just slightly larger than the diameter of the drive axle. Be very careful when gluing the bead to the top plate so that the axle is able to turn freely and maintains vertical alignment. Here is a trick I have used to attach the top bearing:

- Cover the bottom side of the hole with masking tape and then insert the drive axle through the tape. The tape should now be holding the axle centered in the hole (it isn't touching the sides of the hole).
- Place the bead over the axle so that it now rests in the exact alignment it will be in for the engine to work.

- When everything is in perfect alignment, put a few drops of epoxy around the outside of the glass bead so that it will be attached to the pressure chamber top plate. Be careful that the glue does not come close to the inside of the bead or the axle shaft.

**Figure 46 - Use tape to hold the axle in the center of the hole and then carefully glue the glass bead into place.**

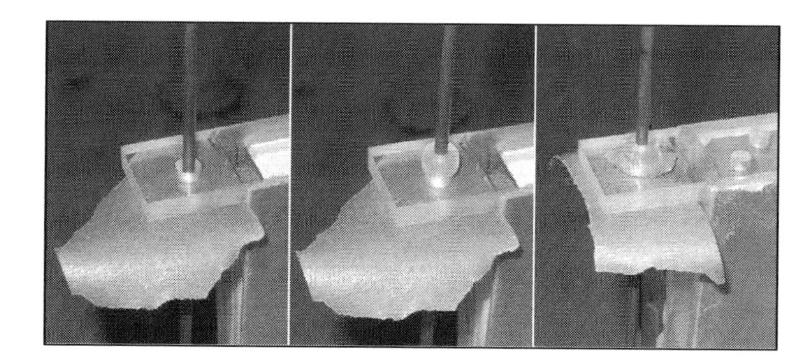

### *Create and Attach Vents*

Venting is important for two reasons. It is occasionally necessary to equalize the pressure inside your motor with the pressure of the environment. If you have done well in creating a sealed pressure chamber then the diaphragm and the attached drive assembly will become very sensitive to small changes in pressure.

The second purpose of the vents is for those who wish to increase the performance of their engine by running it with helium inside. A carefully constructed and well tuned engine of this design will run from the heat of a warm hand in a cool room with just air inside. Adding helium makes a noticeable difference in those situations where the engine does not appear to have enough temperature differential to start running.

Vents are crafted from small fittings made for connecting 1/4" plastic tubing. They can be brass or plastic. Cut the fittings so only the pipe nipple section remains (See figure 24). Use epoxy to glue the nipples over the two small holes on the top plate of the pressure chamber, behind the drive cylinder. I have found I can use the handle of a Q-Tip or a small piece of wire to reach inside the nipple and make sure they are aligned with the holes in the top of the pressure chamber. Remember to use sandpaper to roughen the surface of the acrylic before gluing. Do not let the glue run into the vent hole as this could cause the vent to plug, or it could drip inside and foul up the displacer. I used JB Weld 60 minute epoxy for this joint because it is very thick and sticky, and less likely to run into places where it isn't wanted.

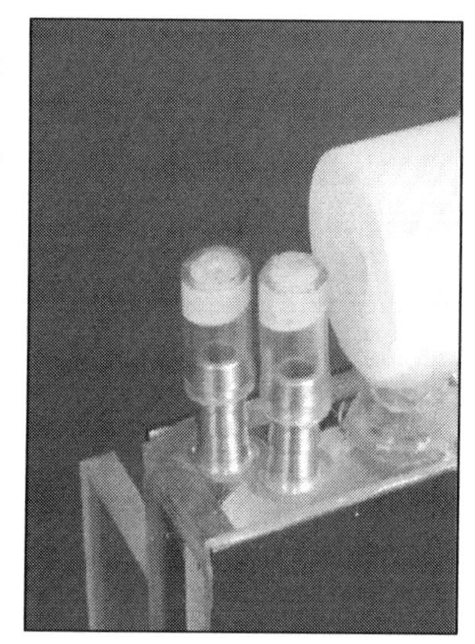

**Figure 47 - Vents are fashioned from 1/4" tubing fittings. Vent plugs are short pieces of tubing sealed with silicone.**

The vent plugs are made from 1" pieces of 1/4" tubing that has had one end filled with silicone glue. Cut two of these and carefully plug one end with silicone. Set them aside until the silicone has cured.

### *Build a Crank Arm*

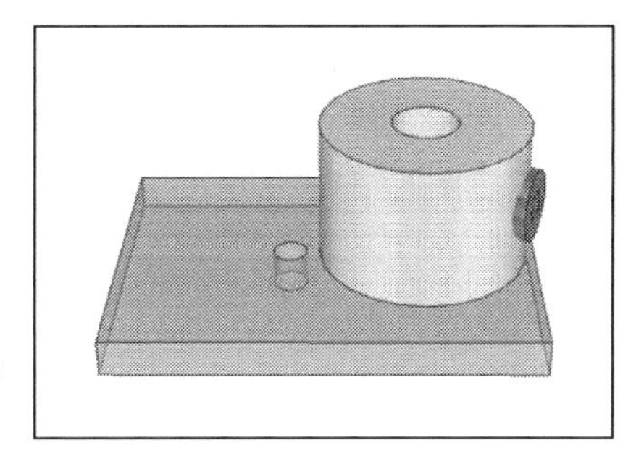

Figure 48 - Crank arm.

The first few crank arms I built were about an inch long and had many holes in them for adjustability. What I have learned is that the only hole I used was the one that was 1/4" from the center.

Cut a piece of thin (3/32") sheet acrylic to about 9/16" wide and about 3/4" long. Use epoxy to glue a 1/8" steel shaft collar near one end of the acrylic piece. Be careful that no glue gets on the set screw. After the glue has cured, back out the set screw so that the 1/8" center hole is not blocked, and drill through the center of the steel collar and through the acrylic with a 1/8" drill bit. The hole will be perfectly aligned if it is drilled after the collar is attached to the acrylic.

A second hole is drilled about 1/4" from the center of the shaft collar. This hole is where the pushrod will connect. I am using 1/16" music wire for my pushrod, so I drilled a 1/16" hole. The hole size must match the material you will be using to make the pushrod. You can see in the illustration that the hole for the pushrod is placed so that it just barely clears the edge of the shaft collar.

### *Cut and Form the Magnetic Drive Arms*

Figure 49 - The drive arms will each hold a magnet. Use a heat gun to bend them into shape. The brackets for the magnets were made separately and glued on.

The magnetic drive is the heart of this design. The magnet drive arms (or "swing arms") ride on the drive axle. They each hold a magnet that is centered on the same axis as the magnet that is in the top of the displacer. As the drive axle rotates it brings a magnet close to the top of the pressure chamber. That magnet is set to repel the magnet in the displacer. When the displacer is moved to the other side the drive axle will stop and begin moving in the opposite direction.

The key to forming these arms is to shape them in such a way that they will hold the magnets in the proper position for moving the displacer.

**Figure 50 - Notice how the drive arms hold the magnets in line with the magnet on the displacer inside the engine.**

Each arm is made from a thin (3/32") piece of acrylic that is 8" long and 1/2" wide. One arm will have the shaft collar glued to the top. The other arm will have the shaft collar glued to the bottom. This will allow the two arms to be close to the same height on the drive axle and will make adjustment and alignment easier to accomplish. Cut these pieces and attach a shaft collar to the center of each one with epoxy. Remember to rough up the surface slightly with sandpaper before gluing. When the glue has cured, drill a 1/8" hole through the acrylic using the collar as a drill guide.

Temporarily mount the arms on the drive axle and put the axle into its place on the engine. The lower arm will have the steel collar on the bottom side of the arm. The upper arm will have the collar on the top. This configuration will keep the arms as close together as possible. Carefully measure the distance the arm will have to be bent to bring the end of the arm in line with the magnet in the displacer.

Use those measurements to make a pattern. Draw the desired magnet arm shape in full scale on a piece of paper. Heat the arm at the point where it needs to be bent and hold it in shape over the pattern until it cools. I have used a variety of swing arm shapes for these engines. The ones pictured here are the simplest and quickest to make. I make one bend in the arm, about 2" from magnet end. I make a small bracket to hold the magnet from a second piece of thin acrylic. The magnet bracket is made from a piece measuring 1/2" x 1", bent in the middle to 90°. This is then glued on with epoxy. Note that the magnet holder is twisted a little to provide the correct angle for the magnet.

Re-mount the arms on the engine and check the alignment of the magnets. They must hold their magnet directly in line with the magnet in the displacer.

Double (and triple) check the polarity of the magnets before gluing them on to the arms. The magnets need to be set to *repel* the displacer. Hold a magnet in your fingers and move it toward the displacer. If the displacer is drawn toward your hand, the alignment is incorrect. Turn the magnet around in your hand so the other pole is facing the displacer and test again. If the displacer pushes away, the alignment is correct.

Mount the magnet on the arm *so that it repels the displacer*. Use sandpaper to rough up the surface of the magnet that will be in contact with the glue. Use epoxy to glue the magnet to the acrylic bracket on the end of the swing arm. The acrylic surface of the swing arm should also be roughed up with sandpaper so the glue will hold.

Do the same for the second arm. It will repel the displacer from the other side.

### *Build the Film Can Drive Cylinder and Attach*

The drive cylinder is fashioned from a 35mm film can and lid. It is connected to the pressure chamber by a short length of vinyl tubing.

Make a hole in the lid of the film can. Remove the material from the center of the lid so that only a circular rim remains. There needs to be only enough material in the lid so that it will still snap back into the canister. This rim will be used to anchor the diaphragm material and hold an air tight seal.

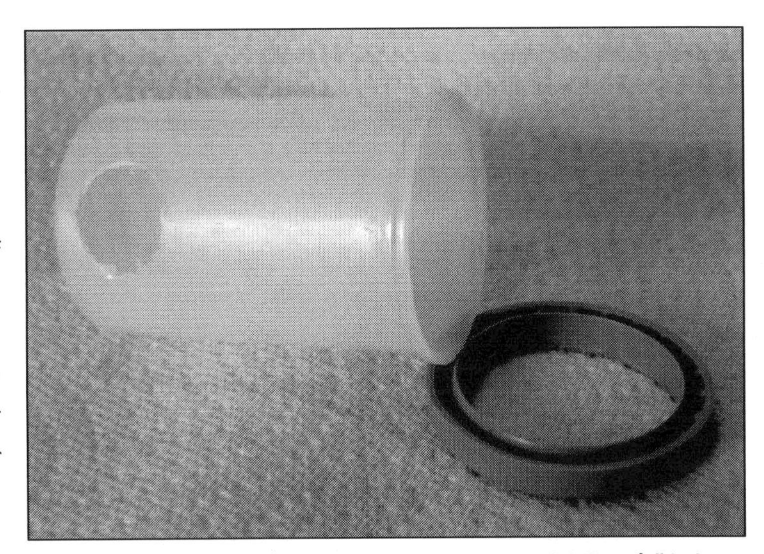

**Figure 51 - Remove the center from the film can lid and drill a 1/2" hole near the base of the can.**

Drill a 1/2" hole in the side of the film can. Center the hole about 5/16" up from the bottom of the canister.

Fashion a small stand from several pieces of acrylic to act as the base for the drive cylinder. The drive cylinder needs to stand about 1/2" above the surface on which it is mounted. It is difficult to get anything to stick to the plastic of the film can, so I have drilled a hole in the base block and tied it to the can with wire. This is in addition to the application of some silicone glue between the two pieces.

You can now cut a short length of 1/2" plastic tubing to connect the film can to the pressure chamber. The tubing needs to be long enough to reach the center of the inside of the film can, and to stick out 3/16" beyond the bottom of the mounting base. This extra protrusion will be used to connect the drive cylinder to the pressure chamber.

Figure 52 - The tubing extends 3/16" beyond the mounting base.

Figure 53 - Looking inside the drive cylinder.

Dry fit the tubing into the film can and into the top of the pressure chamber to make sure everything fits and that the tubing does not interfere with the motion of the displacer. When you are confident that your tubing is the correct length glue it in place with a generous amount of silicone glue. Set it aside until the glue is dry.

The drive cylinder can be glued in place on top of the pressure chamber after the glue has cured. There are two ways to glue it on. If you can get a flush fit between the mounting base of the drive cylinder and the pressure chamber top you can glue it on with acrylic cement using the capillary cementing technique described earlier. If you don't have a great fit, you can glue it on with epoxy. Remember to scuff up any shiny surfaces with sandpaper before using epoxy or it will not stick to the acrylic.

The open end of the drive cylinder should be aimed slightly to the right side of the main drive axle. The pushrod will be attached about 1/4" to the right of center on the drive axle. Aim roughly at that point. The goal is to situate the drive cylinder so that it is pushing straight at its center of effort. Use masking tape to hold the drive cylinder in place until the glue is fully cured.

Figure 54 - The vacuum table is a board with holes in it that is connected to a vacuum cleaner

### *Thermoform the Drive Diaphragm*

Using a shaped diaphragm is necessary for optimal performance. One might assume this is a complicated process, but it is actually quite simple. Thermoforming tools and techniques were outlined earlier in chapter 7. Here are the steps needed to create a thermoformed diaphragm:

- Make a "vacuum table" from a flat board and a vacuum cleaner.
- Make a "form" over which we will mold the vinyl material.
- Mount the vinyl in a frame and heat it.
- Press the warm vinyl over the form and apply a vacuum.

Figure 55 - The "form" is a section cut from a 35mm film can.

The finished diaphragm will be a cylindrical shaped dome that will fit easily inside the film can and is between 1/2" and 7/8" tall.

The "vacuum table" is not really a table at all. It is a board with some 1/8" holes drilled through it, and a vacuum cleaner hose pressed against the back side. The board needs to be situated so it has a level surface on the top and a vacuum hose attached to the bottom side. The picture in chapter 7 shows the board hanging over the side of the table, being held in place by a paint can. The vacuum cleaner hose is held against the table leg with a bungee cord. The hose is situated so that it is in contact with the bottom of the board. When the vacuum is turned on the board gets drawn up snug against the end of the hose and air is drawn in through the 1/8" holes. I usually place a paper towel over the board to create a smooth work surface.

The "form" is the object our vinyl will be molded around. The form is made from a section of plastic film can, the same kind of film can used to make the drive cylinder. Take the lid off of a 35mm film can and mark a line around the outside that is 3/4" down from the open end. Cut carefully along the line. That 3/4" tube will be used as the form for the diaphragm.

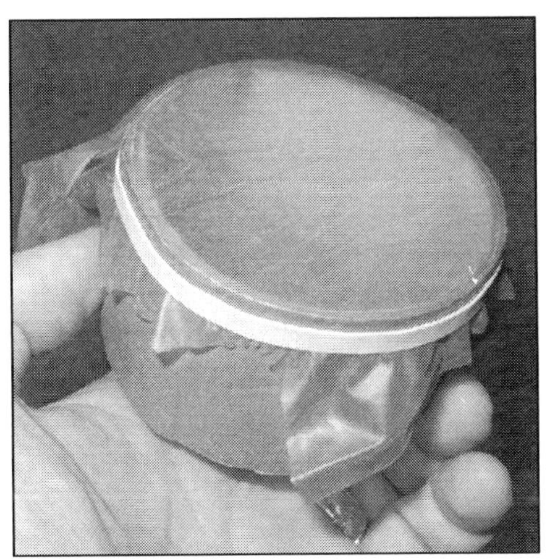

Figure 56 - Vinyl stretched over an old roll of masking tape. It looks like a little drum.

One vinyl glove will provide enough material to make two diaphragms. One can be made from the palm of the glove, and another can be made from the back. Carefully cut the glove into two large flat sheets of vinyl. The fingers of the glove will not be used. They can be cut off and discarded.

A framework of some kind is needed to hold the vinyl flat. I use an empty roll of masking tape and a rubber band as an improvised hoop frame. Place a sheet of vinyl over the end of the old tape roll and hold it in place with a rubber band. Smooth out as many wrinkles as you can. It will look like a little drum.

If you have all those parts in order, and if you have a heat source to warm the vinyl, follow these steps to create the thermoformed diaphragm:

Figure 57 - The warm vinyl is drawn to the shape of the form during thermoforming.

1. Set up the vacuum table. Place a paper towel over the holes to create a smooth work surface. Place the form in the center of the vacuum area. The tube should be placed so that the smooth "factory end" is up, and the edge that you cut is setting on the paper towel. Turn on the vacuum.
2. Stand close to your vacuum table as you heat the vinyl. Use the low setting on the heat gun. Watch the vinyl carefully as it is heated. The material will appear to be wrinkle free and relaxed when it is warm enough for forming. Back off a little with the heat gun when you notice the vinyl is wrinkle free and relaxed, but keep it warm as you move on to the next step.
3. Press the vinyl down over the form and flush to the surface of the vacuum table. The vinyl will instantly conform to the shape of the film can tube. Remove the heat. Turn off the vacuum after about 10 seconds.

Figure 58 - Cut away the excess vinyl. This thermoformed diaphragm will now be a perfect fit with the film can drive cylinder.

If you get the vinyl too hot it will melt and there will be holes in the diaphragm. Try it again with a little less heat. It may take a few tries to learn the technique. After you get one to work, make one or two more. The second and third ones are always easier, and it never hurts to keep a spare diaphragm around.

Remove the vinyl from the frame and carefully cut away the excess material. Carefully coax the thermoformed material into its new shape if necessary.

### *Create and Attach the Piston/Pushrod Assembly*

The next step is to create a lightweight plastic piston and attach it to the diaphragm, and then create an adjustable pushrod that will connect it to the small hole drilled in the crank arm. Here is a picture of the completed assembly.

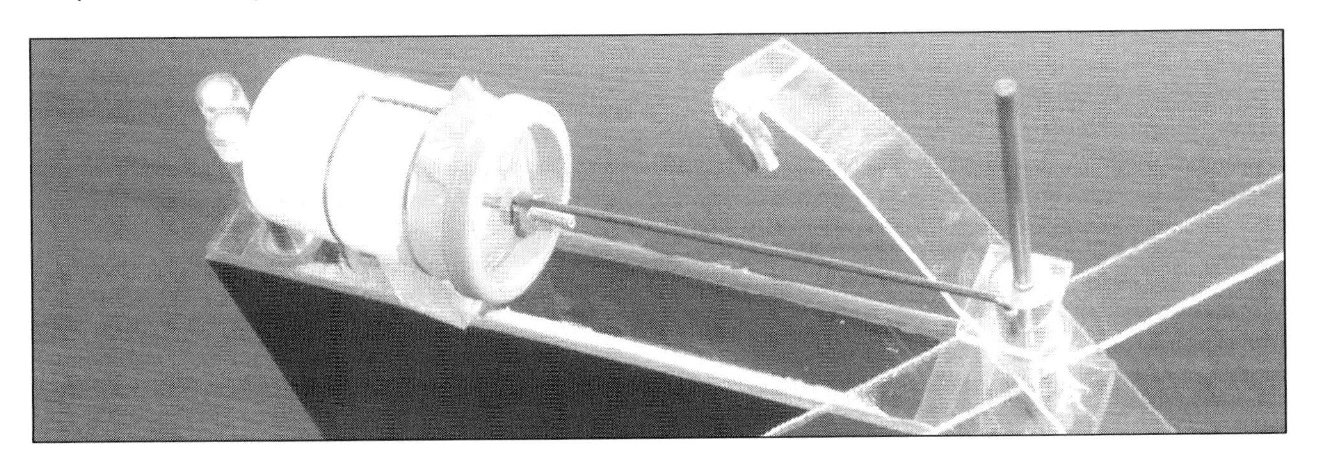

**Figure 59 - There is a lightweight plastic piston attached to the center of the diaphragm which is then attached to an adjustable pushrod that hooks into a small hole in the crank arm.**

### *Piston Disks*

**Figure 60 - Use a quarter to trace and cut two round disks from a disposable coffee cup lid.**

The piston disks are cut from a disposable coffee cup lid. If you don't have one around the house, now would be a good time to go out and get a cup of coffee! Use a quarter to trace two circles on the coffee cup lid and carefully cut them out. Scissors work well for cutting this material. Smooth off any jagged edges so that there are no sharp points to puncture the thin vinyl of the diaphragm. Carefully mark the center of both disks and drill a hole that is the same size as your attachment bolt. Lay one plastic disk on top of the other and drill the holes at the same time. This will guarantee perfect alignment.

If you don't get the holes perfectly centered or aligned the first time, use that as an excuse to go out for another cup of coffee.

### *Attachment/Adjustment Bolt*

The attachment bolt is a 4-40x1" flat slotted machine bolt with three matching nuts. The first nut holds the plastic disks together with the diaphragm material sandwiched in between them. The second and third nuts are used to attach the pushrod.

One disk will be attached to the inside of the diaphragm, and one will be attached to the outside. The attachment bolt will pass through the hole in the center of the disks and hold the entire assembly together. A small drop of silicone glue will seal the hole that is punctured in the center of the diaphragm to prevent any leaks. Here is the process for assembly:

- Place one plastic disk on the inside of the diaphragm. It should be easy to center it in the dome caused by thermoforming.
- Place a small drop of silicone glue on the outside of the diaphragm at the center, where the hole will be.
- Press the second disk onto the outside of the diaphragm and line up the holes in the two disks.
- Use a sharp object such as a needle or a piece of wire to puncture a hole in the diaphragm for the attachment bolt.
- Thread the attachment bolt through the hole from the outside so that the bolt is protruding to the inside of the diaphragm.
- Thread a nut onto the bolt until the two disks are held snuggly together and the silicone is spread sufficiently to form a seal.
- Set the assembly aside until the silicone is cured.

**Figure 61 - A thin plastic disk is bolted to the inside and outside of the diaphragm.**

### *Pushrod*

The pushrod is fashioned from a length of 1/16" music wire. I like the music wire because it is incredibly stiff and will be perfectly straight. Some people may prefer to use a softer wire that is easier to bend, and that is OK. 1/16" craft wire will also work. It is easier to bend into shape but is not as pretty! Start with a piece of wire that is about 7" long. It will be trimmed to a shorter length when finally installed.

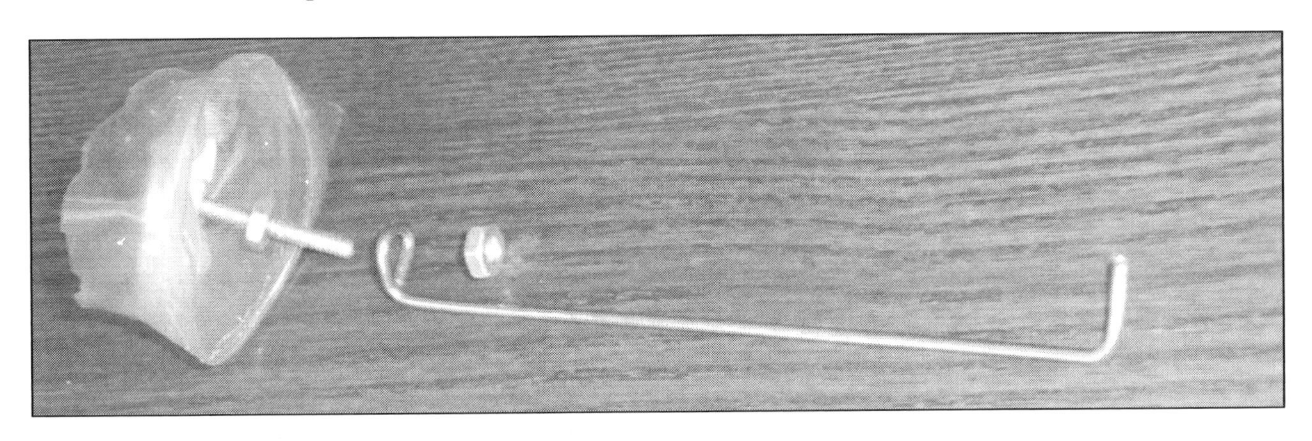

**Figure 62 - Diaphragm with attachment bolt and pushrod. The eye is on the left.**

Use needle nose pliers to create an eye in one end of the pushrod wire just big enough for the bolt to pass through. Then use a heavy set of pliers to bend the eye over to 90°. This eye is used to mount the pushrod to the attachment bolt of the diaphragm/piston assembly. Thread one nut onto the attachment bolt so that it is centered on the exposed bolt. Insert the bolt through the eye of the pushrod and use a second nut to hold it snug against the first nut. This only needs to be finger-tight. If you over tighten this it will be difficult to make adjustments later.

Now do a dry fit of the drive assembly onto your pressure chamber to determine the length of your pushrod. Set your motor on a level surface with the drive axle in place, and with the crank arm attached to the drive axle. Gently insert the diaphragm into the drive cylinder with the pushrod attached. Secure the diaphragm with the film can lid. Measure the pushrod to determine where the bend needs to be so that it will mate properly with the crank arm.

Look carefully at the picture at the beginning of this section that shows the completed assembly. Note that the crank arm is resting to the side of the drive axle. When everything is in "neutral", the crank arm is at a 90° angle from the direction of the pushrod, and the piston is midway between the ends of its normal traveling distance. The pushrod must be cut to the proper length so that the piston can travel freely *inside* the film can. The diaphragm should never be maxed out (pushed in) all the way during the normal operation of the motor. In like manner, the piston disks should never come in contact with the lid of the film can, and they should never be pulled to the outside of the film can during the normal operation of the motor.

From the neutral position you should be able to rotate the drive axle 45° in either direction without bottoming out the diaphragm and without pulling the drive piston out past the lid.

It works best if the vents are open when you are making this measurement.

Work the mechanism back and forth to determine the best working length for the pushrod. Use a permanent marker to mark the spot on the pushrod where it needs to be bent. Remove the diaphragm from the engine assembly and make the final bend in the end of the pushrod. Trim off the excess pushrod so that there is about 1/4" after the bend. Reassemble the diaphragm/pushrod assembly and test it for length. If it is not exactly where it needs to be, use the nuts on the attachment bolt to adjust the length.

### *Build and Attach the Flywheel*

I did a lot of experimenting with flywheels for this engine design, and what I discovered was that the size and weight of the flywheel are not real critical. This engine will operate with a wide variety of flywheel shapes and sizes. It will even run with no flywheel at all.

My first flywheel was made from a CD. The engine ran well but had to be in perfect adjustment to run well with such a light flywheel. I taped coins to the CD for added weight and the engine was easier to start and would run much longer. The added weight made things move slower and farther, and the power stroke of the piston became longer as a result.

And there is more good news. Your flywheel doesn't even have to be round! I like to make the flywheel from a round disk of clear acrylic because it looks so nice. But quite frankly it would work just as well if it was a square or a triangle as long as it was well balanced and about the same weight. If you are using simple hand tools the round acrylic flywheel is a lot of work to make. It works just as well if it is made of wood.

But for those who wish to make a clear round flywheel, here are the steps to make it:

- The finished flywheel should be approximately 6" in diameter, 1/4" thick, and have a 1/8" inch hole in the center.
- Cut a piece of 1/4" acrylic sheet to just over 6" square. Do not remove the protective coating.
- Mark the center of the disk and drill the 1/8" hole. Use a drill bit that is approved for acrylic material.

**Figure 63 - If you have a band saw you can make a simple circle cutting jig to make a round flywheel.**

- Make a circle cutting jig for the band saw. A circle cutting jig in this case is a large scrap of wood with a small finish nail driven into it. The board is clamped to the band saw table so that the nail is 3 inches to the right of the blade.
- Start cutting the circle free-handed until you can align the hole in the acrylic with the pin in the jig. When the acrylic is on the pin rotate the material slowly in a clockwise direction until you have cut a perfect circle.
- Remove the disk from the jig. Do not remove the protective coating. Begin sanding the edge of the disk with 100 grit sandpaper. Use several finer grades until you finally finish with 400 grit or 600 grit.
- Polish the edge with plastic polish or rubbing compound.
- Now you can remove the protective coating.
- Attach a 1/8" shaft collar to the center of the disk, aligned with the 1/8" hole that was drilled earlier.

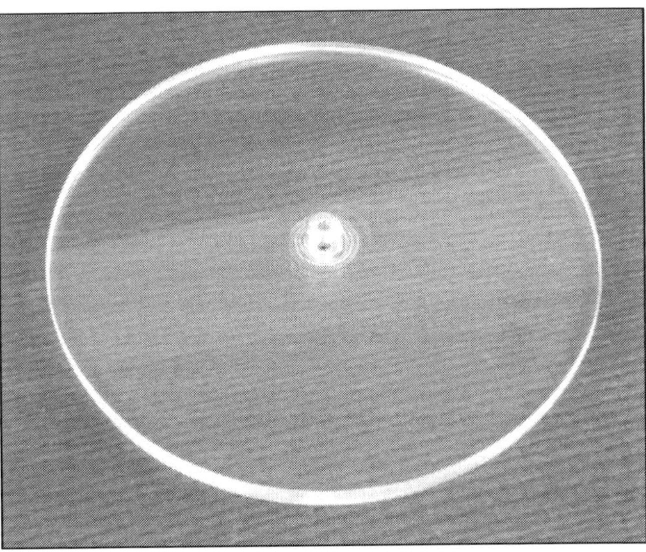

**Figure 64 - 6" Acrylic flywheel with a 1/8" steel shaft collar attached at the center.**

The flywheel is the last item attached to the drive assembly. It rests on the top of the drive axle above the swing arms for the magnets. My wife tells me that it makes my engines look like the starship Enterprise.

### *Pre-Flight Checklist*

Let's just double check to make sure everything is in order. It is time to re-attach any parts you have been taking on and off during assembly so that we can start your engine for the first running!

You should now have an assembled Stirling engine with the following components:

- A sealed pressure chamber mounted vertically on a stand. The aluminum plating on the sides is painted black.
- The displacer inside the pressure chamber can rock easily from one side to the other. If you swing the magnetic swing arm close to the top of the pressure chamber it will push the displacer to the opposite side.
- The drive axle is held at the top in a glass bead bearing, and the rounded bottom end of the drive axle is resting on the smooth part of a penny. Mounted on the drive axle are the magnetic swing arms, the crank arm, and the flywheel.
- The drive cylinder is attached to the top of the pressure chamber and the tube that connects it to the pressure chamber has an air tight seal.

- The drive assembly (diaphragm, piston, pushrod, and crank arm) are installed and adjusted correctly.
  - The crank arm is in its neutral position at 90° to the body of the engine and the piston pushrod.
  - The crank arm is mounted level with the center of the drive cylinder.
  - The diaphragm has an air tight seal to the film can.
  - The diaphragm needs to be set so that the piston can freely travel at least 1/8" in either direction out of the neutral position. Adjust the pushrod length to allow for proper piston movement.
- Close the vents when the piston is in neutral position.

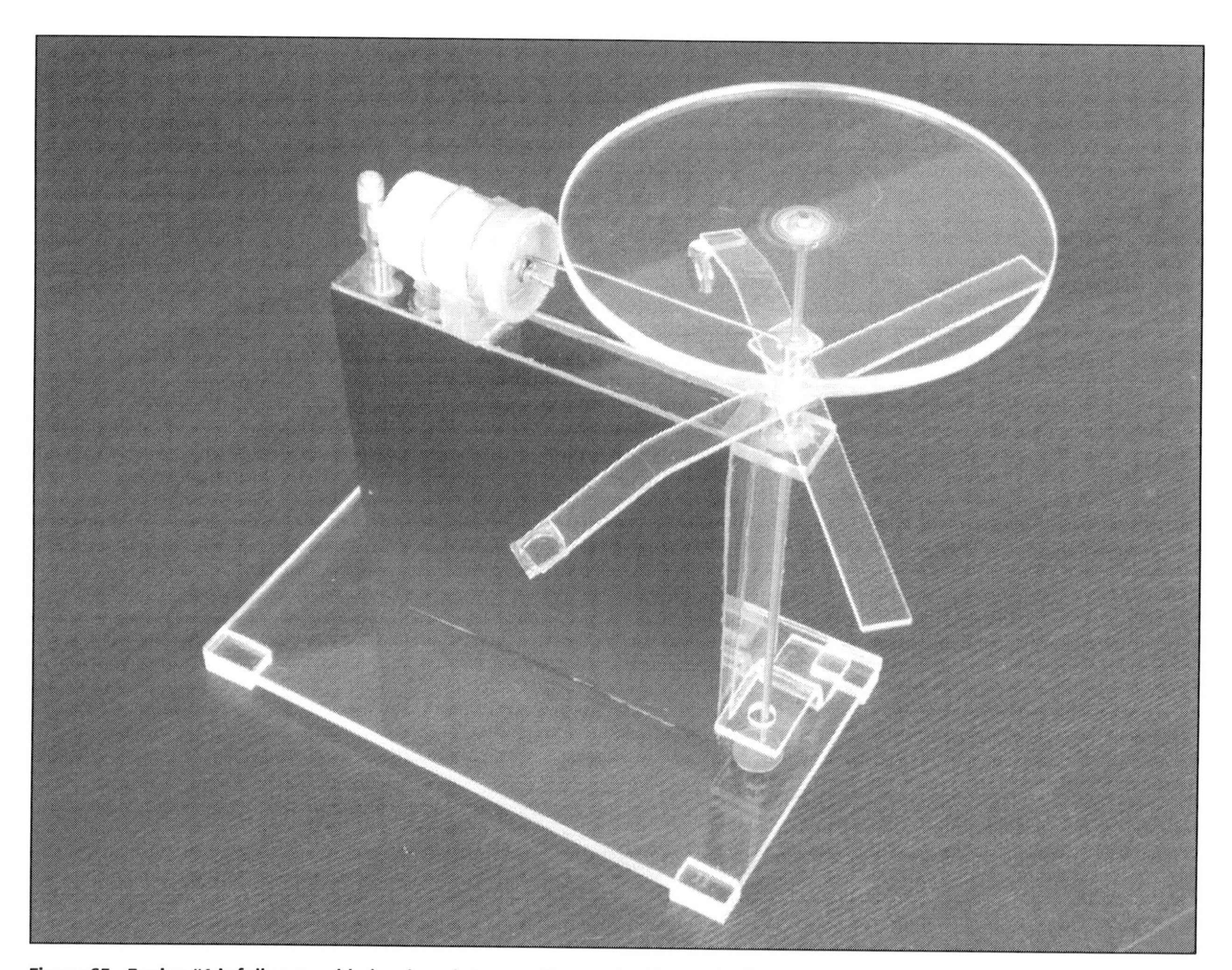

**Figure 65 - Engine #1 is fully assembled and ready to run. You are looking at the "warm side" of the pressure chamber.**

## Operation and Fine Tuning

Check for pressure leaks. The easiest way to do this is to note the position of the diaphragm when sitting idle. Next, pull out on the pushrod for a few seconds and then release. The diaphragm should return to the same position it was in before the test. If it does not return to the same position there may be a pressure leak. Check all the joints and seal up any leaks if they exist.

A good starting position for the magnetic drive arms is to cross them at about 90° and make sure they are evenly distanced from the magnet in the displacer. Rotate the drive assembly back and forth a few times to see that the action on each side of the displacer is about equal.

I will sometimes lubricate the drive axle at the point where it passes through the pressure chamber top plate and at the point where it touches the penny. Spray a small amount of silicone lubricant onto a Q-tip and dab a small amount on the axle at these two points.

The pushrod and diaphragm need to be adjusted so that they do not restrict the movement of the flywheel mechanism. Rotate the drive axle back and forth so that you can observe the displacer move back and forth in both directions. The drive piston needs to be able to travel freely without being restricted by the diaphragm material. If you feel resistance coming from the diaphragm (because it gets too tight, or because the piston hangs up on the cylinder) then adjust the pushrod length until there is no resistance felt.

If you have assembled your engine as illustrated, the "warm side" is the side that does *not* have the stand braces. Another indicator is that the warm side of the engine is the same side that the crank arm is on. This engine will not run if the temperature differentials are reversed. You must always apply heat to the warm side. This side of the engine is active when the displacer is on the other side of the chamber. Air expands when it is on the warm side, pushes on the diaphragm, rotates the drive axle and eventually moves the displacer to cover the warm side plate.

While it is true this engine will run from the heat of a warm hand, this is not the most rewarding thing to try first. That exercise will come later. The easy way to get your engine started is to shine a 25 to 60 watt incandescent bulb a few inches from the warm side of the engine. If you need an additional energy boost during your initial tuning, place a bag of frozen peas on the cold side of the engine. This will increase the temperature differential and help you get started with the tune-up process. After you have some time to observe and adjust your drive mechanism you will be able to run the engine without any ice.

The engine will warm up in 1 to 2 minutes and will need a slight push to get started. If it does not start running right away, allow another minute or two for the engine to warm up.

I mentioned earlier I invested about $20 in a small infrared thermometer. I can point it at the surface of the engine and get an immediate temperature reading. This engine should idle along well for you the first time with a temperature differential of 30° to 40°. After you do some fine tuning it should consistently start and run with a temperature differential of 30° or less. When fine tuning is completed this engine will run with a temperature differential of 20° or less. The engine you see in the pictures here will run down to a temperature differential of 12° with air inside. If I fill the pressure chamber with helium it will run on a temperature differential as low as 9°.

I should point out that I have built several of these and I do not get the exact same performance every time. I have pointed out many times that there are variables in the building process that can have a big impact on your final results. You must build with great care to achieve the 12 degree temperature differential operation.

I need to also point out the way temperature differentials are measured. They are measured on the surface of the running engine, not on the source of the heat. Right now my hand is 93.3°. The room temperature is 70°. But when I use my hand to run the engine in this environment I can warm up one side to about 86°, and the temperature of the other side will rise to about 75°. That is just shy of the 12° needed to make it run. In order to run my engine by the heat of my hand in here today I would need to drop the room temperature a few degrees or add helium to the pressure chamber.

Here are the adjustments you can make as you tune your engine for peak performance:

1. Adjust the length of the pushrod on the drive piston to change the relative position of the power stroke within the cylinder.
2. Open a vent to equalize the pressure after the engine has warmed up. Close the vent with the drive piston in neutral position.
3. Move both magnet arms so that the magnets are closer to the pressure chamber to shorten the throw of the engine. (Short stokes make the engine work well with smaller pressure differentials.)
4. Move both magnet arms so that the magnets are farther away from the pressure chamber to lengthen the throw of the engine. (Long throws make the engine work better with larger pressure differentials.)
5. Move one or both magnet arms in order to center them around the pressure chamber.
6. Rotate the crank arm slightly to re-center the drive arms around the pressure chamber.

The trick to tuning is to observe the engine carefully and make one adjustment at a time. If it improves performance, keep it. If it doesn't help, change it back. If it degrades performance, try changing it the other way. I always watch the movement of the diaphragm as I tune the engine. It is usually easy to tell what the engine is doing by watching the pressure changes on the diaphragm.

My favorite way to run these engines is to set them in the sunshine. On a 70 degree day in the Pacific Northwest the sun will heat the warm side of the engine to as much as 120°, providing a 50° temperature differential. This works better than artificial heat sources, such as an incandescent light.

### Adding Helium

There are 2 vents in case you want to try running your engine with helium. I recommend you start using your engine with air. This will help you get a feel for how to tune it and make it work. Then add helium to see the performance difference that it makes. To add helium, tip the engine so that the vents are on the bottom edge of the pressure chamber. Open both vents. Connect one vent to a helium balloon and discharge about 1/2 of its contents into the engine, then seal the pressure chamber by closing both vents. This process has now flushed most of the room air out of the engine and replaced it with the lighter helium.

When running on air or helium it is sometimes necessary to open one vent briefly so that the interior of the engine can equalize pressure with the environment. Changes in the weather bring changes in barometric pressure. Your pressure chamber is a crude barometer and is sensitive to these atmospheric pressure changes. If you are careful you can equalize the pressure several times before you notice a loss of performance caused by a loss of helium.

***If your engine won't start, take a look at some of these trouble shooting tips:***

| Symptom | Possible Cause | Solution |
|---|---|---|
| Engine will not start running | Temperature differential is too low. | Increase temperature differential by adding heat to the warm side or ice to the cool side. |
| | There may be a pressure leak. | Check all joints, seams, vents, etc. and seal any leaks. |
| | There may be too much friction. | Check all bearing points. Make sure the penny is in place. Apply minute amounts of silicone based lubricant if needed. Check the diaphragm assembly to make sure it is not hanging up on something. |
| | Heat is on the wrong side. | Heat the same side of the engine the pushrod is mounted on. |
| Displacer spends most of its time on one side, and only briefly moves to the other side. | Magnetic drive alignment problem. | Move the magnet farther away from the side that is getting only brief contact. You may need to briefly open a vent to reset the piston to its neutral position. |
| Displacer only cycles slowly a few times and then stops. | The magnets may be too close. | Move both magnets farther away from the displacer. |
| The engine starts but does not run for more than a few cycles. | The flywheel may be too light. | Tape several coins (evenly spaced for balance) to the flywheel. |
| The piston does not move far enough to make the displacer move. | Magnets are too far away or too weak. | Move magnets closer to the displacer or try stronger magnets. |

| | | |
|---|---|---|
| | The crank arm setting is too long. | Shorten the crank arm if you can. |
| | There may be excess pressure inside the pressure chamber. | Open a vent, move the piston to the neutral position, and close the vent. |

With a little fine tuning the engine will run on the small light bulb or sunshine with very little intervention from you.

### Running from the Heat of Your Hand

Remember that the engine does not run on heat alone. It runs on a temperature differential. There must be heat, and there must be something cooler. If you are in a very warm room, or if it is a hot summer day, the environment will not be cooler than the palm of your hand. It is also true that not all people have warm hands.

It is easiest to start and run one of these engines from the heat of your hand when the engine is cool, and when it is in a cool environment (below 70°). Place the engine on a level surface. Rest the palm of your hand against the warm side of the engine. Pay close attention to the drive mechanism and you will see it move slightly as the engine warms up. After a minute or so, open one of the vents and re-center the drive mechanism, then close the vent.

Give the engine a little push to get it started. If it doesn't start right away you may need to continue warming for a little longer. If your hand starts to get cool, turn the engine around so you can place your other hand on the warm side and continue warming the engine. It should start with a little encouragement from you in about another minute.

The magnet drive arms usually have to be situated a little closer to the pressure chamber when running from the heat of your hand. After a little practice you will learn how to set up your engine for running with a variety of different temperature differentials.

This process is much easier when helium has been added to the pressure chamber. An engine filled with helium will run in a warmer room because it does not require as great of a pressure differential.

A helium charge seems to last from 7 to 10 days. After that time the performance seems to return to the room-air baseline.

# Chapter 10: Engine #2 - Horizontal Flywheel Magnetic Drive Stirling Engine

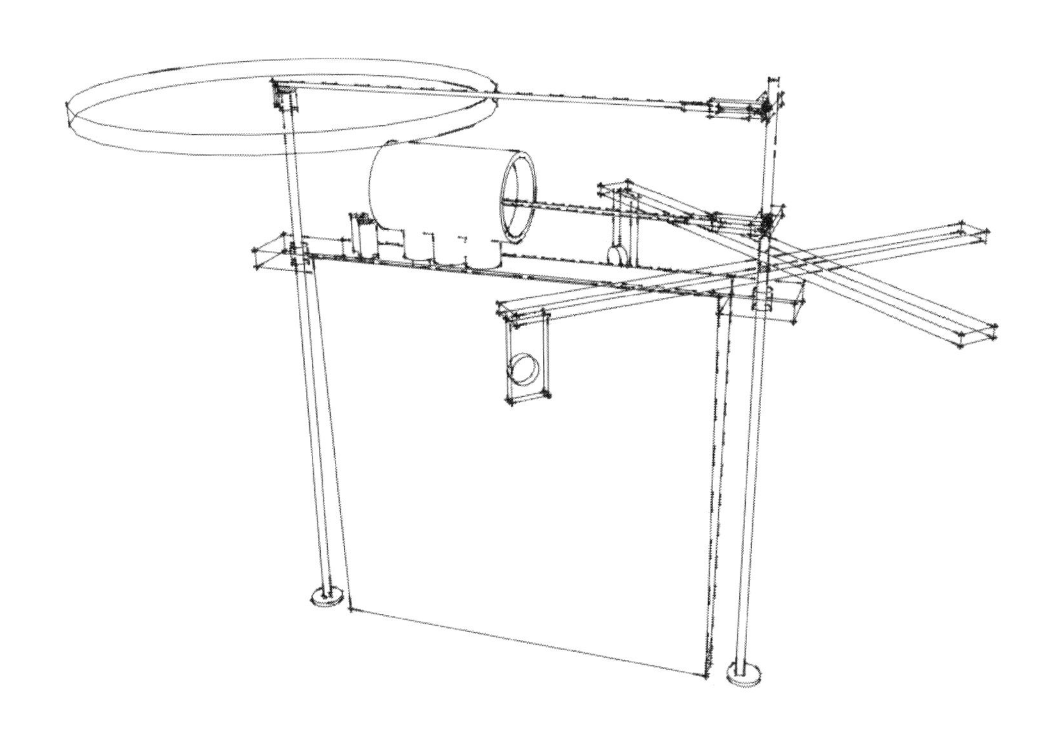

## Engine Design Explained

Engine number two is very similar in design to the first engine. It is in fact the same engine with the addition of a rotating flywheel. The flywheel increases the engine's ability to store inertial energy by continuously rotating in one direction. However the addition of the flywheel eliminates the self adjusting feature of the first engine. This engine will not have a variable displacement as temperature differentials rise and fall because the displacement is held steady by the flywheel crankshaft. You will find it is very picky about some adjustments.

The engine runs on a temperature differential by making one side of the engine warmer than the other. The displacer (a piece of poster board that is loose inside) moves the air from the warm side to the cool side and back again. The air expands and contracts in volume as it heats and cools.

The expanding and contracting air causes the piston to move out and in repeatedly. This out and in motion rocks the first part of the drive mechanism back and forth, and also serves to move the displacer back and forth.

The second pushrod takes the rocking action of the first part of the drive mechanism and converts it to the rotating motion of the flywheel.

The engine retains the vertically oriented magnetic drive displacer. The advantage of this part of the design is that it eliminates the need for a gland and pushrod arrangement of traditional Stirling engines which can cause friction or leaks (especially when made to the low tech standards required for these projects). The vertical orientation allows for a rocking motion that moves relatively easily and does not have to be counter balanced.

The horizontal orientation of the flywheel makes it possible to use a simple bearing arrangement. Once again the axle will be shaped to a rounded point and will ride on a hard surface.

There are two glass bead bearings in this design. These are simple decorative glass beads from a craft store that happen to fit the 1/8" shaft that is used in the engine design. The glass bead provides a low friction surface and has good wear characteristics.

This engine has two shafts, which adds several friction points to the entire design. The additional axle carries the rotating flywheel. The benefits of adding a rotating flywheel balances out the added friction of the flywheel mechanism.

### *Pardon Me for Repeating Myself*

As you read through the instructions for Engine #2 you will notice a lot of instructions are very familiar and sound virtually identical to the instructions for Engine #1. I weighed the options and decided to write up this section of the book so it could stand alone for those who just wanted to build this design without having to constantly refer back to the previous chapter for instructions. I have eliminated some of the anecdotal comments to make it a little shorter for the reader.

## Parts List

The table below represents the retail prices in March of 2010. Many of the supplies I purchased came in large quantities that will provide enough material to make several engines. Because of that, I have listed 2 prices. The first column is the price I paid for the material. The second column represents the adjusted cost for the amount of materials that were actually used in this engine.

**Engine #2**

| Item | Count | Retail | Actual |
|---|---|---|---|
| 6" x 12" x 0.064 Aluminum Sheet | 1 | $ 7.29 | $ 7.29 |
| 1/8" (0.125) Music Wire | 1 | $ 1.16 | $ 0.58 |
| 1/16" (0.062) Music Wire | 1 | $ 1.88 | $ 0.94 |
| 3/32" (0.093) Clear Acrylic | 1 | $ 2.79 | $ 0.20 |
| 1/4" (0.220) Clear Acrylic | 1 | $ 14.97 | $ 3.74 |
| 1/4" Brass Barb Splicer | 1 | $ 2.18 | $ 2.18 |
| Vinyl Gloves (10 pk) | 1 | $ 3.97 | $ 0.40 |
| 1/8" Steel Shaft Collar | 6 | $ 1.00 | $ 6.00 |
| 1/4" ID Vinyl Tubing (10') | 1 | $ 3.11 | $ 0.05 |
| 1/2" OD Vinyl Tubing (10') | 1 | $ 4.15 | $ 0.03 |
| 35mm Film Can | 2 | $ - | $ - |
| Neodymium Magnets 1/16" x 1/2" | 4 | $ 0.45 | $ 1.80 |
| JB Weld Epoxy | 1 | $ 4.97 | $ 0.99 |
| 5 Minute Epoxy | 1 | $ 3.97 | $ 0.79 |
| Clear Silicone | 1 | $ 3.89 | $ 0.97 |
| 3/16" Foam Poster Board | 1 | $ 3.62 | $ 0.08 |
| Black Spray Paint | 1 | $ 0.97 | $ 0.24 |
| Helium Filled Balloon | 1 | $ - | $ - |
| Penny | 1 | $ 0.02 | $ 0.02 |
| 4-40 x 1 Machine Screws and Nuts (10pk) | 1 | $ 0.99 | $ 0.33 |
| Acrylic Cement | 1 | $ 5.30 | $ 0.53 |
| 1/8" ID Glass Bead | 2 | $ 0.10 | $ 0.20 |
| | | $ 66.78 | $ 27.37 |

Since I am a bargain hunter, I will always shop around for the best prices. If you are lucky enough to find scrap acrylic for $1.00 a pound and a good discount hardware store you should be able to bring that $66.78 figure down to about $46.00.

## Conceptual Drawings

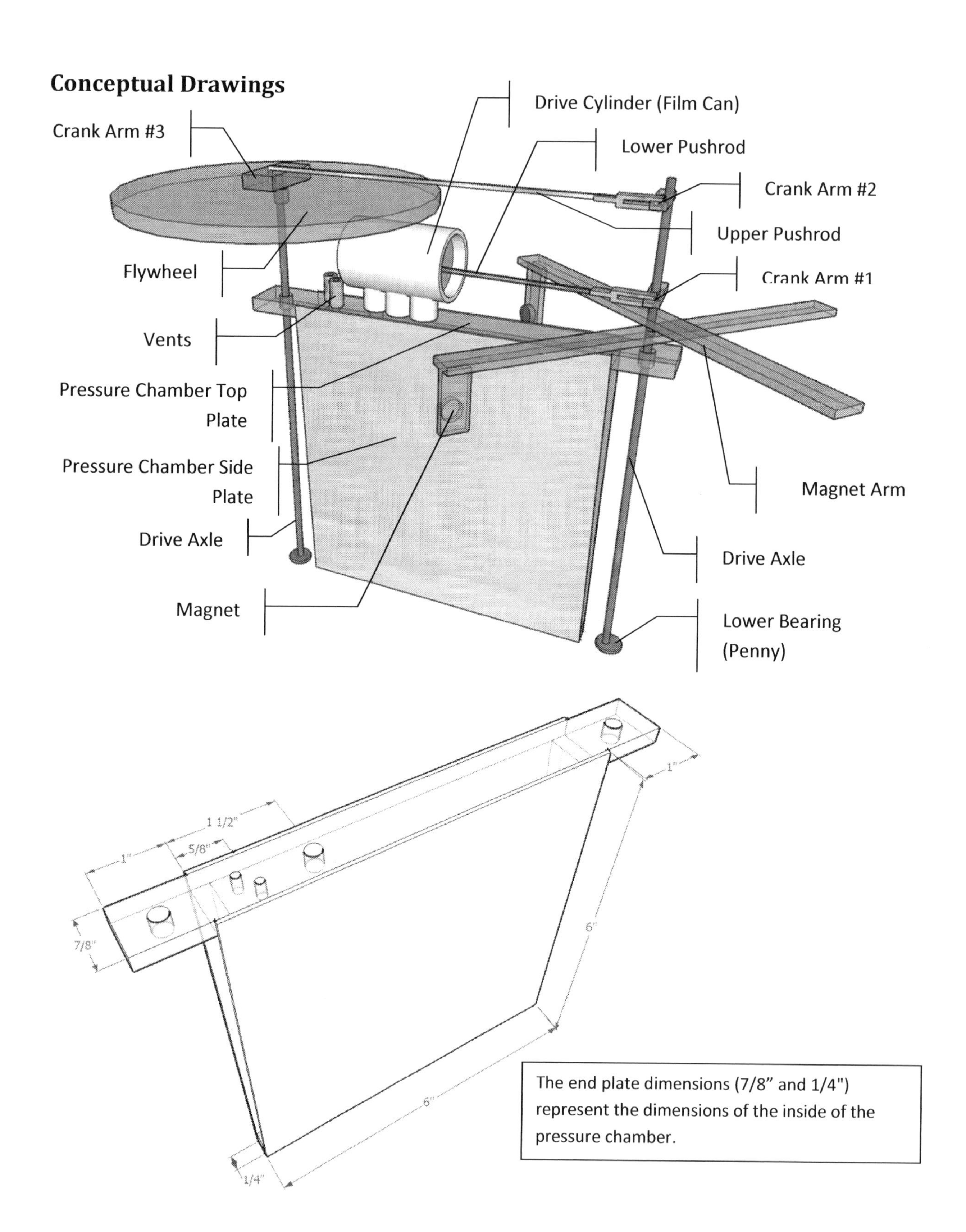


The end plate dimensions (7/8" and 1/4") represent the dimensions of the inside of the pressure chamber.

**And here is another tip:** You will notice as you compare the first two designs that the pressure chambers are almost identical. Engine #1 has a single tab protruding from one side at the top to hold the drive axle. Engine #2 has a tab at both ends to hold two axles. The top acrylic piece is the only difference between these two pressure chambers. If you build the pressure chamber for Engine #2 you can set it up to run as a reciprocating engine (Engine #1) or as a flywheel driven engine (Engine #2). The second bearing is the only difference in the pressure chamber design. The two designs also share similar drive cylinders and magnetic swing arms. So if you want to try both designs and want to avoid building two pressure chambers, start with this design first.

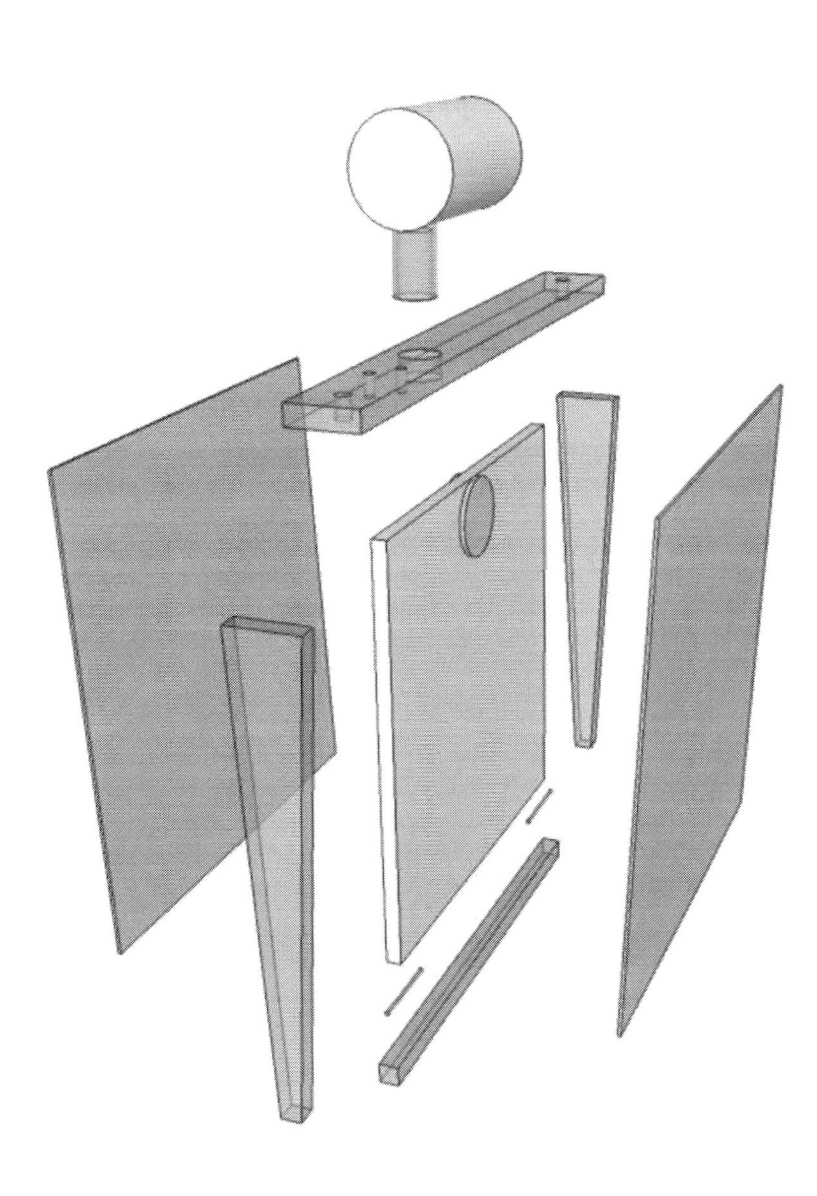

## Step by Step Instructions

### *Draw Plans and Create Templates*

Here are the dimensions of the pressure chamber parts. You will find it very helpful if you take some time and draw your own full scale plans and templates before you begin. The templates are particularly useful for creating the acrylic parts of the pressure chamber. The flywheel (not shown here) is a 6" disk of 1/4" acrylic. Keep that in mind as you lay out your parts for cutting. The holes for the axles are 3/16". The vent holes are 1/8". The hole for the drive cylinder is 1/2". It is intended to be a snug fit for the tubing used to connect the drive cylinder.

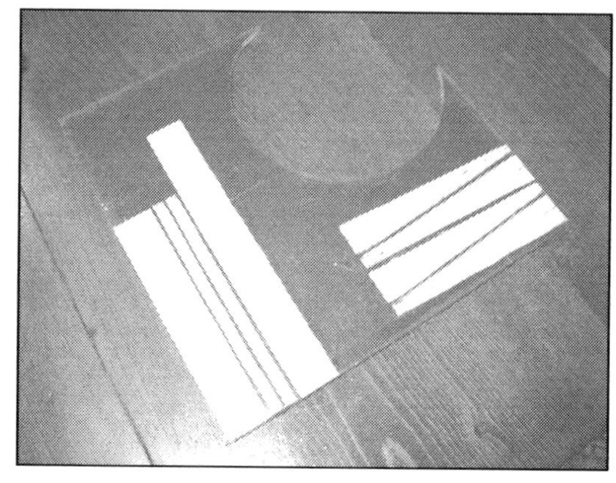

**Figure 66 - The templates for two pressure chambers are laid out on the acrylic sheet in preparation for cutting. Making full sized templates is very useful in this process.**

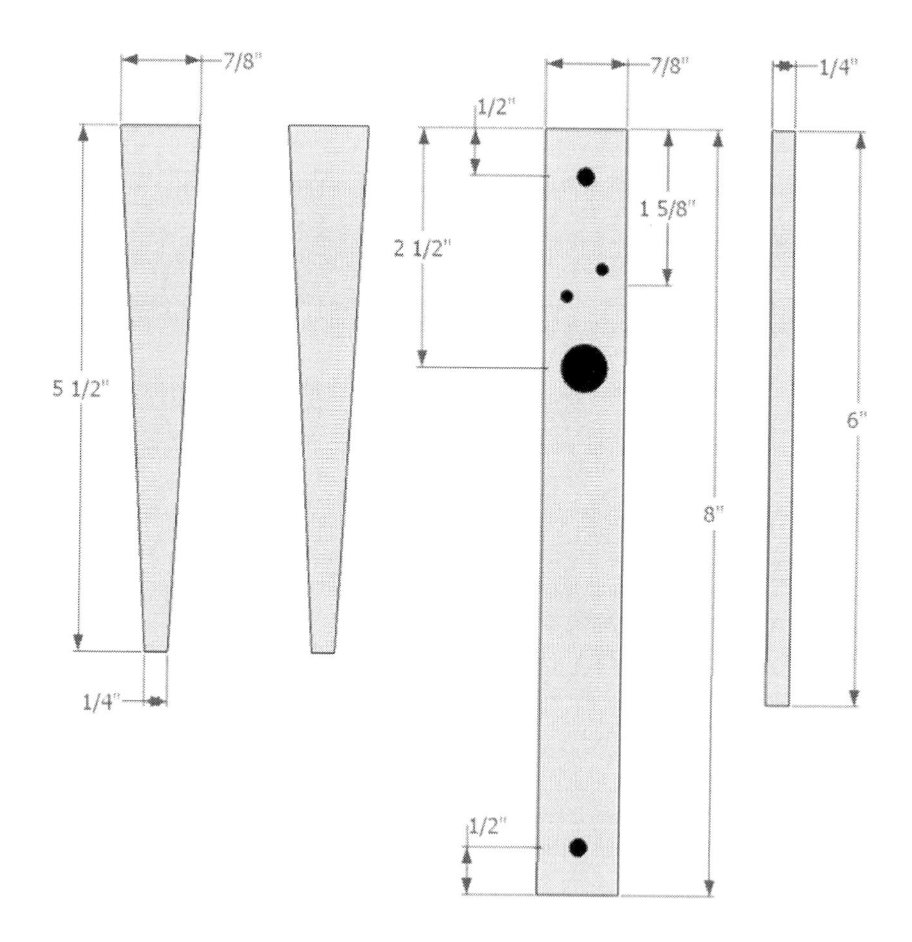

### *Cut and Paint the Aluminum for the Pressure Chamber Side Walls*

Figure 67 - Pressure chamber side walls are painted black on both sides except for the inside border (shown here).

You will need to cut two 6" squares of aluminum sheet. The aluminum I purchased for this project comes in a sheet that measures 6" by 12". Only one simple cut was required.

The aluminum must be of a heavy enough gauge so that it will not easily flex. Do not use thin flimsy metal as it will not perform as well. The material here has a thickness of 0.064. Aluminum that has a thickness near 0.04" will also work. If it is much thinner than that it becomes too flexible for a pressure chamber.

I don't recommend the use of tin snips for cutting flat sheets of metal. The tin snips will bend the metal on one side of the cut.

You can cut this by hand with a hack saw if you are patient, but it will be a challenge because a hack saw is not really intended for cutting material that is 6" wide.

I recommend either cutting it with a nibbler or with a metal cutting blade in a saber saw. If you use the nibbler you will need to flatten the edge of the metal after it has been cut, but that is easy to do. If you use a saber saw you will want to use a board and some clamps to hold the metal securely to your workbench and then use the board as a cutting guide. Always secure metal parts with clamps before attempting a cut with a power tool. You don't want these sharp objects flying through the air at high speed! It can get quite noisy, so hearing protection is recommended.

Please refer to the chapter, "Working with Aluminum" for a detailed explanation of cutting and finishing techniques.

Use masking tape to prevent painting the area around the inside edge of each sheet where the acrylic parts will be attached to the aluminum. This will help provide a strong glue joint. Glue sticks better to bare aluminum than it does to painted aluminum. Paint both sides of the aluminum sheets black.

The black paint really does make a difference in engine performance. I have made Stirling engines with black paint, white paint, and no paint. I now paint all sides of my aluminum sidewalls with flat black spray paint.

### *Cut the Acrylic Parts for the Pressure Chamber Bottom, Ends, and Top*

Figure 68 - Pressure chamber top, bottom, and end pieces cut from 1/4 inch acrylic sheet.

The pressure chamber dimensions are made to accommodate the material chosen for the displacer. My displacer material is about 3/16" thick. If you are able to locate material for your displacer that is 3/16" thick then just use the dimensions provided in the drawings at the beginning of the chapter. If for some reason you cannot find an exact match for the displacer material you must adjust the dimensions of your pressure chamber accordingly. i.e., If your displacer is 1/4" thick then you should make the pressure chamber 1/16" wider at both the top and the bottom.

Lay out the full scale pattern for each of the acrylic parts and attach it to the surface of the acrylic for cutting. Do not remove the protective coating from the acrylic sheet until after the parts are cut. Leave enough space between the parts to accommodate the width of the cut that will be caused by your saw. I used a saber saw for cutting the parts you see in these illustrations.

Use a sanding block if necessary to make all cut surfaces in the acrylic flat and straight. Both end pieces (the tapered ones) must be identical. Tape or clamp these pieces together when sanding the edges in order to maintain a consistent shape. Only sand the cut edge. Do not sand the shiny flat surface of the acrylic. If cutting leaves a buildup of melted acrylic you can sometimes scrape it off. Use another piece of acrylic as a scraper to remove the rough particles that may be stuck to the cut edge.

The long edges of the top plate and the bottom plate should be angled to match the angle of the end pieces. If you don't have the ability to cut this angle reliably, cut them square and then sand them to the correct angle. It is possible to sand the edges after the frame is cemented together by rubbing the completed frame on a sheet of flat sandpaper.

Drill the holes in the pressure chamber top plate prior to assembly. The holes are less likely to cause a crack or a break in the acrylic if you drill them before cutting. The holes for the bearings are 3/16". The vent holes are 1/8". The hole for the drive cylinder is 1/2"

### *Assemble the Pressure Chamber – Leaving One Side Off*

You may begin to assemble the pressure chamber when the paint is cured on the aluminum parts and the acrylic framework pieces have been cut to size.

Figure 69 - Hold the acrylic pieces so that nothing touches the joint as you join the parts.

Dry assemble all parts first to guarantee a good fit. Shape or replace any parts that are not fitting well.

Assemble the acrylic frame using the capillary cementing technique described in chapter 6. This provides an attractive and sturdy joint for the acrylic frame. When the frame is complete and the joints have cured you will attach it to one of the aluminum side plates. (The other aluminum plate will be glued on after the displacer is assembled and placed inside.)

I learned the hard way that you don't want any clamps or braces (or even tape) to be touching the acrylic joint when joining with solvent. The solvent will run under the tape or the clamp and disfigure your beautiful acrylic surface. The picture here shows how I held the parts square while cementing them together so that nothing would be touching the joints when the solvent was applied.

I recommend using a slow cure epoxy like JB Weld for attaching the aluminum side plates to the acrylic framework. The slow cure will provide you with plenty of time to work with the joint so that you won't be rushed by a five minute time limit like you would with faster curing glue. Also, JB Weld is very thick and sticky, which is helpful for this operation. You want to make sure you don't get too much glue oozing out of the joint. Large globs of glue on the inside of the pressure chamber can interfere with the operation of the displacer after the engine is assembled.

Figure 70 – Engine #2 pressure chamber frame assembly is glued to one aluminum side plate with epoxy. One side is left off to enable installation of the displacer.

You will not be able to wipe the epoxy off the acrylic if you make a mistake. If you dribble a little glue on your acrylic, and it is not in a place that will interfere with the operation of moving parts, my advice is you leave it there. Trying to clean it off will only make it look worse.

### *Build the Displacer*

The Displacer will be cut from 3/16" thick poster board. Two thin neodymium magnets will be attached to the center of the top edge, and two small wires will be attached at the bottom. The wires will extend slightly past the end of the displacer and will prevent the displacer from getting close enough to touch the acrylic end plate. The wire also provides a smaller footprint for the displacer and reduces friction that would be caused by having the entire edge of the displacer on the bottom of the pressure chamber.

The dimensions of the displacer will vary slightly depending upon the thickness of the acrylic used to make the end plates. The displacer should be cut so that there is about 1/8" clearance on all sides when it is inside the pressure chamber.

**Figure 71 - Small wires are attached to the bottom edge of the displacer. The tape is removed after the wires are glued in place.**

Measure the inside height and width of the pressure chamber. Cut the poster board to be 1/4" shorter than the inside of the pressure chamber (this will make 1/8" clearance on each side).

Check the fit and mark which side of the displacer will be at the top. Attach a thin (1/16") neodymium magnet to each side of the displacer, centered along the top edge. My magnets were 1/2" round and 1/16" thick. Use silicone or epoxy to glue the magnet to the outside top edge (centered). The magnets must be installed so that they are attracting each other. This is very convenient as it makes it possible to glue them to the displacer without a clamp. The direction of the poles does not matter as long as the magnets are set to attract.

I have also built these with a single magnet imbedded inside the foam core of the displacer material. You can do it this way if you can't find thin magnets. You should understand that the magnets serve a second purpose when mounted to the outside edge of the displacer. They prevent the displacer material from resting flush against the aluminum sidewall. If you choose to place your magnet inside the foam core you will have to add a small spacer (such as a drop of glue) to the top edge of the displacer so that it will always stand off from the side wall about 1/16".

Cut two pieces of thin wire to a length of 1". Use sandpaper or a file to remove any burrs from the cutting process. Use tape to temporarily attach one to each bottom corner of the displacer. They should be taped to the end opposite the magnet, so that they rest flat on the bottom of the pressure chamber when assembled.

The ends of the wire should protrude 1/16" on each side. This will act as a "stop" to prevent the displacer from getting too far out of center.

Figure 72 - Displacer is in position and the displacer is checked for size. The magnet is at the top (wide side) and the two wires are at the base (narrow side).

Gently place the displacer assembly inside the pressure chamber to inspect the fit. If both of the short wires are touching the sides of the frame at the same time they are too long. The displacer must be able to rock back and forth with no obstruction. Hold the second side plate in place and check to make sure the displacer clears all sides by about 1/8" as it moves back and forth.

When you are happy with the fit, use epoxy to glue the wires in place on the bottom edge of the displacer. Use tape to hold the wires in place and set it aside until the glue is completely cured, then remove the tape.

### *Attach the Second Pressure Chamber Side Plate*

This is one of the more complicated and tricky parts of this assembly process. The displacer must be placed inside the pressure chamber so that the magnets are at the top (the wide end) and the wires are at the bottom (narrow end). The side plate must be glued in place to make an air tight seal, but still allow the displacer to move freely from side to side.

If glue oozes from the joint and sticks to the displacer, and the displacer can't move freely, the engine will not run.

**Here is how I do it:**

- Place the partially assembled pressure chamber on a flat surface with the open side facing up.
- Place the displacer inside the pressure chamber.
- Place the second aluminum side plate in place (with no glue yet) and inspect the fit. It should be a flush tight joint.
- Remove the side plate and apply a thin coat of JB Weld 60 minute epoxy to the acrylic frame side of the joint.
- Press the aluminum side plate against the glue and apply enough pressure to squeeze the excess glue from the joint.

- Immediately remove the aluminum side plate and inspect the glue joint to make sure you have an air tight seal. You should see a continuous trail of wet glue all the way around the aluminum side plate. If there is excess glue oozing into the interior of the pressure chamber remove it now. Be careful you do not accidentally glue the displacer to the inside of the pressure chamber.
- Replace the side plate and gently clamp it with a rubber band or weighted object until the glue is dry.
- Keep the pressure chamber flat until the glue is cured. Do not pick it up or try to make the displacer move while the glue is still sticky. If you cause your displacer to move while there is still tacky glue inside it could glue your displacer to the inside of your engine and your engine won't work.

### *Build a Stand for the Pressure Chamber*

At this point you should have an assembled pressure chamber that has a clear acrylic frame (top, bottom, and ends), two black aluminum side plates, and a displacer. The displacer should move freely from side to side, and you should be able to make it move by holding another magnet near one of the magnets on the displacer. It is now time to build some type of apparatus that will hold it upright and will hold the bottom bearing for the drive axle.

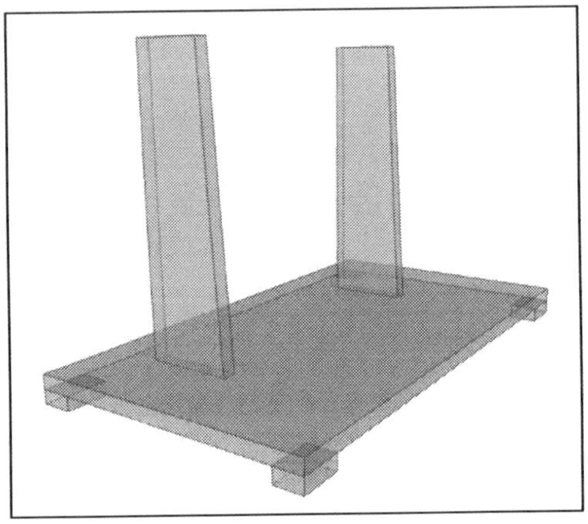

**Figure 73 - The stand holds the pressure chamber upright.**

The base can be built of wood or acrylic. Acrylic is only slightly harder to work with and makes a very attractive stand. I have used both materials and favor the acrylic for its good looks. The stand needs to provide a wide base for the assembled engine so that it does not fall over. It needs to be at least one inch longer than the pressure chamber. For this engine I used a scrap of 1/4" acrylic sheet that measures 5" by 9". I cemented four small scraps of acrylic to the bottom of the base to act as feet. These are intended to prevent the bottom of the acrylic base from getting scratched.

The two vertical bars are approximately 1 1/2" by 5", also made from scraps. One of the long sides on each vertical brace is cut at an angle so that the pressure chamber will stand with its centerline perfectly straight up and down. The vertical braces must be set as close to the edge of the pressure chamber as possible. Set them so that the outside edge of the brace is flush with the end of the pressure chamber.

You can find the angle for the vertical brace by tracing the same template you used to cut the pressure chamber end plates. You can use a carpenter's bevel tool to check your angles prior to final assembly. If the centerline of the pressure chamber is vertical, the angle will be the same on both sides.

### *Attach the Pressure Chamber to the Base*

The cool side of the pressure chamber will rest against the vertical braces of the stand. Later when you assemble the drive mechanism the pushrod will be attached on the warm side of the engine.

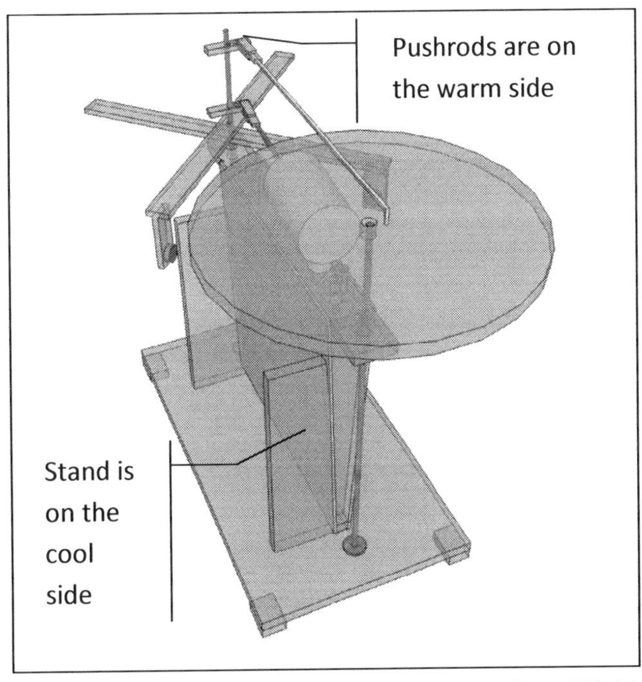


Figure 74 - Build a stand with angled vertical braces that will hold the pressure chamber with its centerline perfectly vertical. The stand is attached to the cool side of the engine. Both pushrods are attached on the warm side of the engine.

Use clear silicone to attach the pressure chamber to the base. Apply a small bead of silicone adhesive along the contact area of the base and the pressure chamber. Check the angles on both sides of the pressure chamber and when they are correct, use tape or a rubber band to hold the pressure chamber in place until the silicone has set.

Silicone is used because it is not necessarily permanent. If you ever need to disassemble this joint to correct the angle you will be glad you used silicone.

**Note:** The centerline of the pressure chamber must be perfectly vertical! Both sides of the pressure chamber must slope at the same angle. This is critical to the balance of the engine. If these angles do not match, the displacer will not move easily in both directions. Have I said this before?

### *Cut and Shape Two Drive Axles*

Cut two 9 3/4" lengths of 1/8" music wire. The top ends should be finished smooth with a file or a sanding block so that all the burrs are removed. Use a file or sanding block to sharpen the lower ends of the axles so that they resemble a dull pencil. The goal is to make a small smooth tip on which the drive assembly will ride. A small smooth point will reduce the friction at the spot where the axle rides on the copper penny. If the tip is too sharp it will bore a hole into the penny and increase friction. If you don't sharpen the tip of the shaft you will experience more friction as a result of the increased contact area between the end of the shaft and the coin.

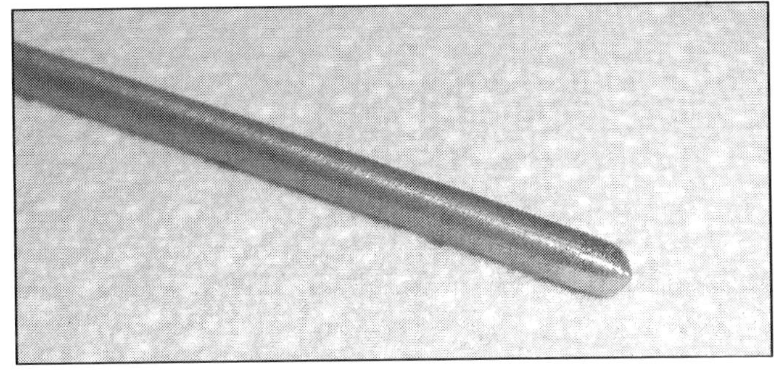

Figure 75 - The bottom end of the axles are sharpened to a dull point to reduce friction.

### *Create the Drive Axle Bearings*

Figure 76 - The keeper bracket does not touch the drive axle, but acts as a safety to prevent misalignment. The drive axle rests on a penny. There is bearing and a keeper bracket on each end of the pressure chamber.

The lower bearings consist of two pieces: A small bracket to prevent the axle from getting out of alignment, and a penny on which it rides.

Prepare the penny by putting a thin coat of silicone on the back side of the penny and let it cure. After it cures, the silicone will act as a non-skid surface and prevent the coin from wandering while the engine is running.

The bracket is not intended to touch the drive axle. It is only there to prevent the axle from moving out of alignment, which would damage the top bearing. The bracket seen in the photos was bent from a small piece of 3/32" acrylic, and the hole was drilled using a carbide bit. The bracket has to allow the penny to be placed under the end of the axle. The axle will ride on one of the flat smooth areas surrounding Lincoln's head.

I used a heat gun to warm the brackets, and pressed them into shape with some scraps of wood. They can be glued to the base by a capillary cementing process, with silicone, or with epoxy. If you use epoxy to glue acrylic you must rough up the surfaces with sandpaper first, otherwise the glue will not hold.

Figure 77 – Acrylic sheet is easily heated and bent to form the keeper for the bottom of the drive axles.

The top bearing is a glass bead. The hole in the bead is just slightly larger than the diameter of the drive axle. Be very careful when gluing the bead to the top plate so that the axle is able to turn freely and maintains vertical alignment. Here is a trick I have used to attach the top bearing:

- Cover the bottom side of the hole with masking tape and then insert the drive axle through the tape. The tape should now be holding the axle centered in the hole (it isn't touching the sides of the hole).
- Place the bead over the axle so that it now rests in the exact alignment it will be in for the engine to work.
- When everything is in perfect alignment, put a few drops of epoxy around the outside of the glass bead so that it will be attached to the pressure chamber top plate. Be careful the glue does not come close to the inside of the bead or the axle shaft.

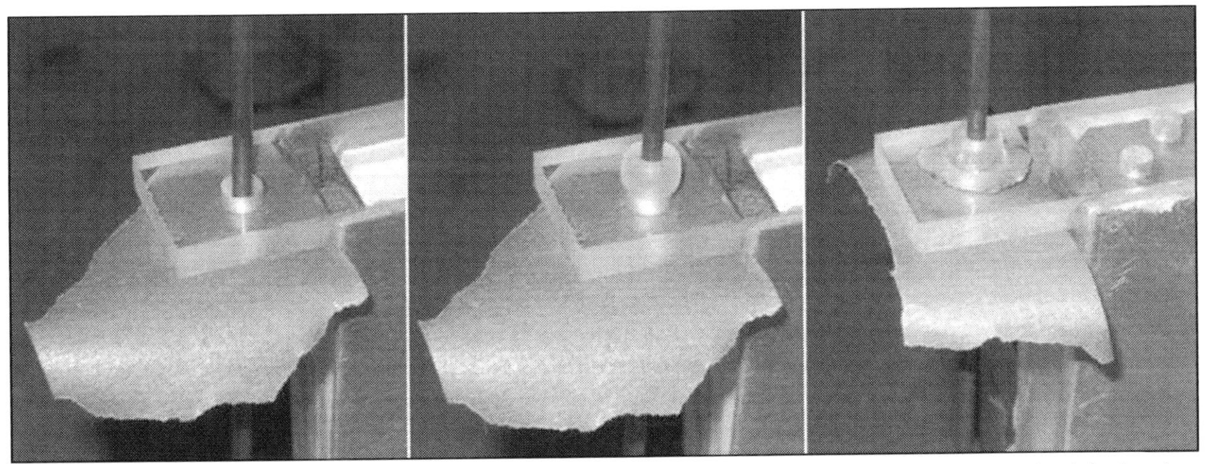

**Figure 78 – Center the shaft in the hole with tape. Set the bearing over the shaft. Epoxy around the outside of the bearing.**

### *Create and Attach Vents*

Venting is important for two reasons. It is occasionally necessary to equalize the pressure inside your motor with the pressure of the environment. If you have done well in creating a sealed pressure chamber, the diaphragm and the attached drive assembly will become very sensitive to small changes in pressure.

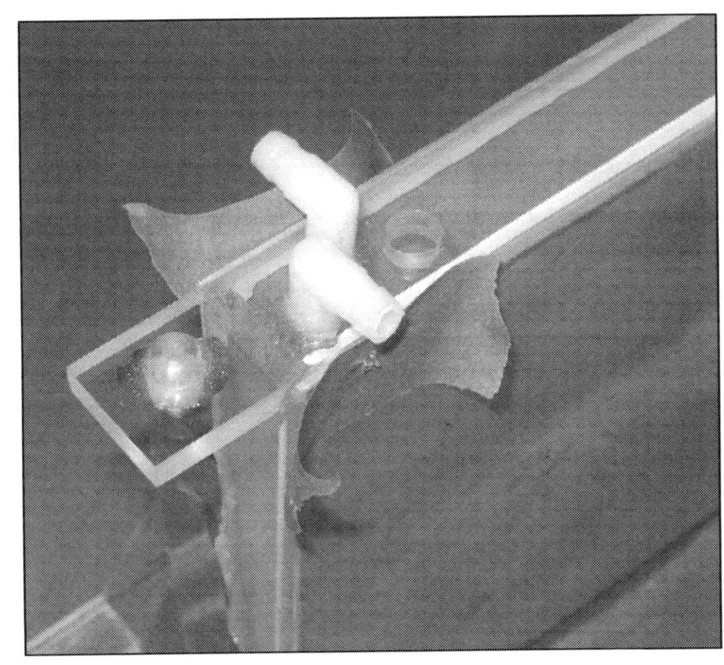

**Figure 79 - Vents are fashioned from 1/4" tubing elbows. These are held in place with tape as the glue cures.**

The second purpose of the vents is for those who wish to increase the performance of their engine by running it with helium inside. A carefully constructed and well tuned engine of this design will run from the heat of a warm hand in a cool room with just air inside. Adding helium makes a noticeable difference in those situations where the engine does not appear to have enough temperature differential to start running.

The vents are crafted from small fittings made for connecting 1/4" plastic tubing. They can be brass or plastic. Cut the fittings so that only the pipe nipple section remains (See figure 24). Use epoxy to glue the nipples over the two small holes on the top plate of the pressure chamber, behind the drive cylinder. As you

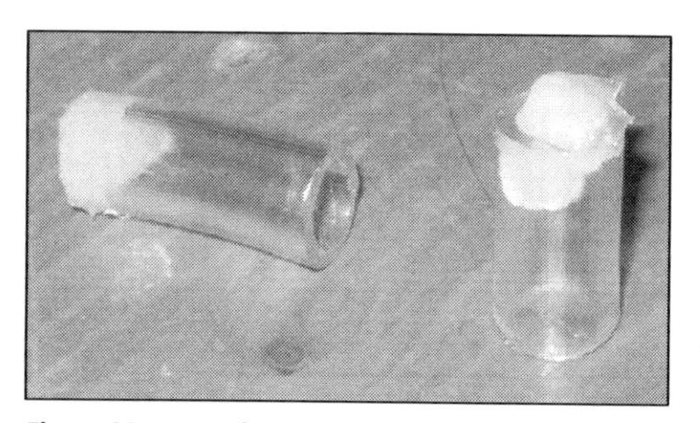

Figure 80 - Vent plugs are made from short pieces of tubing that have had one end filled with silicone glue.

can see in the illustration here, I sometimes use straight nipples and sometimes use ones that are angled. Remember to use sandpaper to roughen the surface of the acrylic before gluing. Do not let the glue run into the vent hole as this could cause the vent to plug, or it could drip inside and foul up the displacer. Use epoxy for gluing the vents in place.

The vent plugs are made from 1" pieces of 1/4" tubing that has had one end filled with silicone glue. Cut two of these and carefully plug one end with silicone. Set them aside until the silicone has cured.

### *Build Three Crank Arms*

This engine design requires that you have the ability to change some adjustments to accommodate for different running environments. I accommodate this by making cranks that have multiple holes in them. This lets you quickly make adjustments to your engine.

There are three crank arms that need to be made. The first two are made from 3/32" acrylic and will have holes drilled all the way through them. These cranks will reciprocate back and forth just like the crank arm on Engine #1.

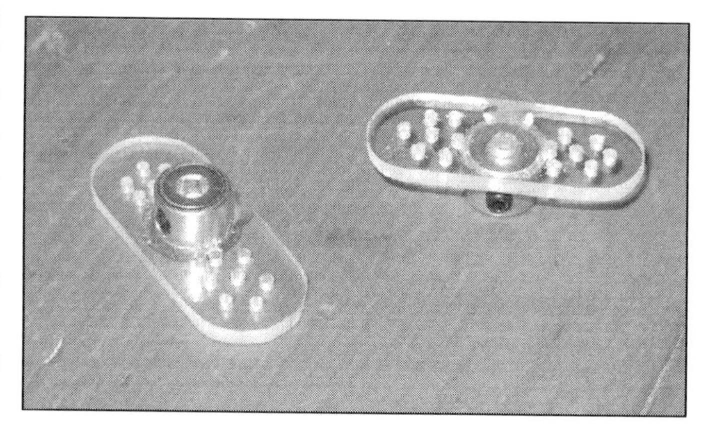

Figure 81 - Build two crank arms from 3/32" acrylic and steel shaft collars.

To make these crank arms cut two pieces of 3/32" acrylic to 1 1/2" by 9/16". Use epoxy to glue a 1/8" steel shaft collar to the center of each piece. After the glue has cured, back out the set screw and gently drill a 1/8" hole through the center, using the shaft collar as a drill guide. Drill an assortment of 1/16" holes on both ends of the crank arm. Make at least one of the holes as close as you can to the shaft collar.

The third crank arm is made from 1/4" acrylic. The holes in this crank arm do not go all the way through. This allows the pushrod to rest in the bottom of the hole and maintains clearance as the flywheel rotates. If the holes are drilled all the way through this piece the pushrod will rub on the surface and possibly get hung up.

To make the third crank arm cut one piece of 1/4" acrylic to 1 1/2" by 9/16". Use a 1/16" drill bit to drill a series of holes to a depth of approximately 3/16" (about 3/4 of the way through). Space the holes about

3/32" apart so that you are able to drill about 10 holes in the entire piece. This side will be the top of the crank arm.

Use epoxy to glue a shaft collar to the center on the bottom of the piece. Do NOT drill out the center of this crank arm.

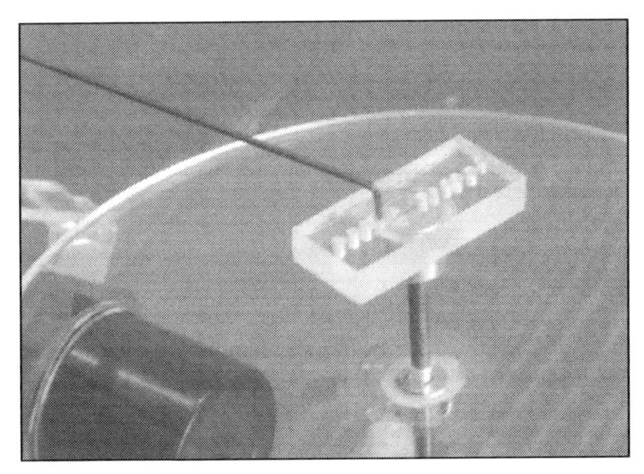

**Figure 82 - Build one crank arm from 1/4" acrylic. Drill holes in the top of the crank arm about 3/4 of the way through the acrylic.**

### *Cut and Form the Magnetic Drive Arms*

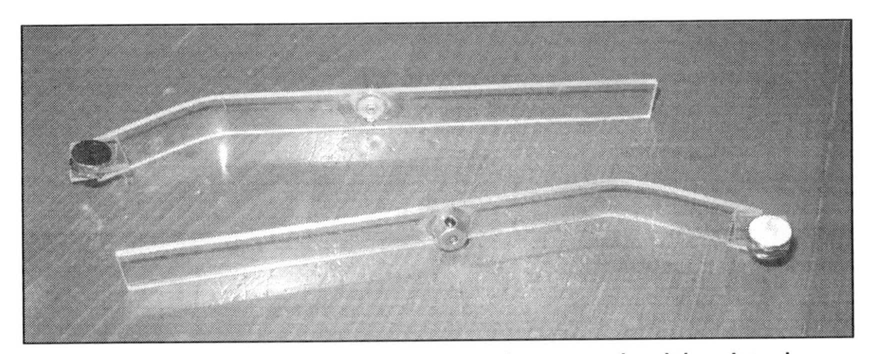

**Figure 83 - Drive arms each hold a magnet. Use a heat gun to bend them into shape. The brackets for the magnets were made separately and glued on.**

The magnetic drive is the heart of this design. The magnet drive arms ride on the drive axle. They each hold a magnet that is centered on the same axis as the magnet that is in the top of the displacer. As the drive axle rotates it brings a magnet close to the top of the pressure chamber. That magnet is set to repel the magnet in the displacer. When the displacer is moved to the other side it will cause the drive axle to stop and begin moving in the opposite direction.

The key to forming these arms is to shape them in such a way that they will hold the magnets in the proper position for moving the displacer.

Each arm is made from a thin (3/32") piece of acrylic that is 8" long and 1/2" wide. One arm will have the shaft collar glued to the top. The other arm will have the shaft collar glued to the bottom. This will allow the two arms to be close to the same height on the drive axle and will make adjustment and alignment easier to accomplish. Cut these pieces and attach a shaft collar to the center of each one with epoxy. Remember to rough up the surface slightly with sandpaper before gluing. When the glue has cured, drill a 1/8" hole through the acrylic using the collar as a drill guide.

Temporarily mount the arms on the drive axle and put the axle into its place on the engine. The lower arm will have the steel collar on the bottom side of the arm. The upper arm will have the collar is on the top. This configuration will keep the arms as close together as possible. Carefully measure the distance the arm will have to be bent to bring the end of the arm in line with the magnet in the displacer.

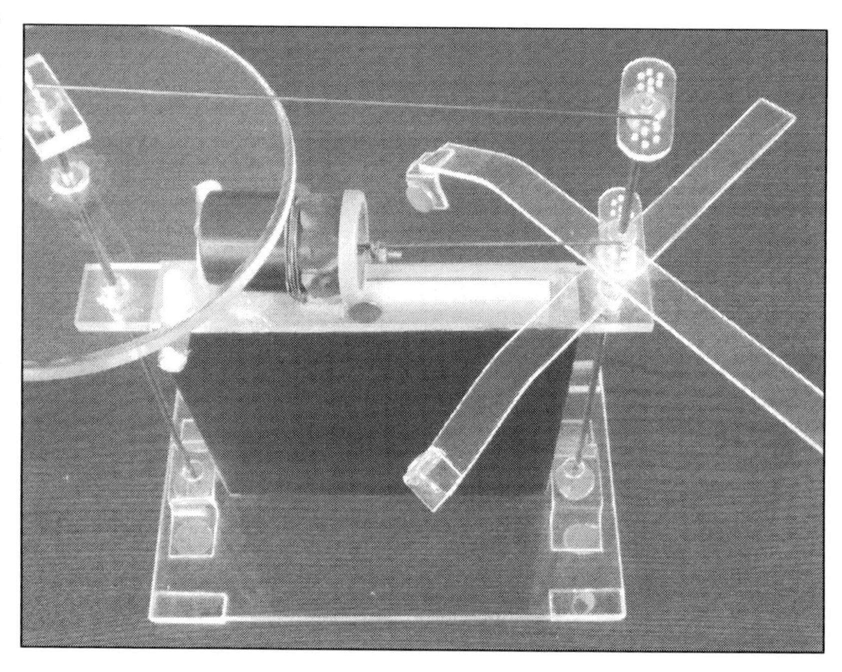

Figure 84 - Notice how the drive arms hold the magnets in line with the magnet on the displacer inside the engine.

Use those measurements to make a pattern. Draw the desired magnet arm shape in full scale on a piece of paper. Heat the arm at the point where it needs to be bent and hold it in shape over the pattern until it cools. I have used a variety of swing arm shapes for these engines. The ones pictured here are the simplest and quickest to make. I make one bend in the arm, about 2" from magnet end. I make a small bracket to hold the magnet from a second piece of thin acrylic. The magnet bracket is made from a piece measuring 1/2" x 1", bent in the middle to 90°. This is then glued on with epoxy. Note that the magnet holder is twisted a little to provide the correct angle for the magnet.

Re-mount the arms on the engine and check the alignment of the magnets. They must hold their magnet directly in line with the magnet in the displacer.

Double (and triple) check the alignment of your magnets before you glue them on to the arms. The magnets need to be set to *repel* the displacer. Hold the magnet in your fingers and move it toward the displacer. If the displacer is drawn toward your hand, the alignment is incorrect. Turn the magnet around in your hand so the other pole is facing the displacer and test again. If the displacer pushes away, the alignment is correct. Mount the magnet on the arm so that it repels the displacer. You can mount the magnets with silicone or epoxy.

Do the same for the second arm. It will repel the displacer from the other side.

### *Build the Film Can Drive Cylinder and Attach*

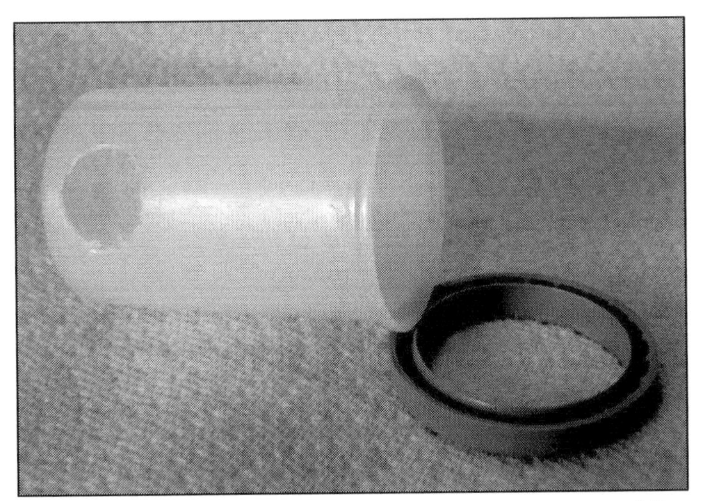

Figure 85 - Remove the center from the film can lid and drill a 1/2" hole near the base of the can.

The drive cylinder is fashioned from a 35mm film can and lid. It is connected to the pressure chamber by a short length of vinyl tubing.

Make a hole in the lid of the film can. Remove the material from the center of the lid so that only a circular rim remains. There needs to be only enough material in the lid so that it will still snap back into the canister. This rim will be used to anchor the diaphragm material and hold an air tight seal.

Drill a 1/2" hole in the side of the film can. Center the hole about 5/16" up from the bottom of the canister.

Fashion a small stand from several pieces of acrylic to act as the base for the drive cylinder. The drive cylinder needs to stand about 1/2" above the surface on which it is mounted. It is difficult to get anything to stick to the plastic of the film can, so I have drilled a hole in the base block and tied it to the can with wire. This is in addition to the application of some silicone glue between the two pieces.

You can now cut a short length of 1/2" plastic tubing to connect the film can to the pressure chamber. The tubing needs to be long enough to reach in to about the center of the inside of the film can, and to stick out 3/16" beyond the bottom of the mounting base. This extra protrusion will be used to connect the drive cylinder to the pressure chamber.

Dry fit the tubing into the film can and into the top of the pressure chamber to make sure everything fits and that the tubing does not interfere with the motion of the displacer. When you are confident that your tubing is the correct length glue it in place with a generous amount of silicone glue. Set it aside until the glue is dry.

Figure 86 - The tubing extends 3/16" beyond the mounting base.

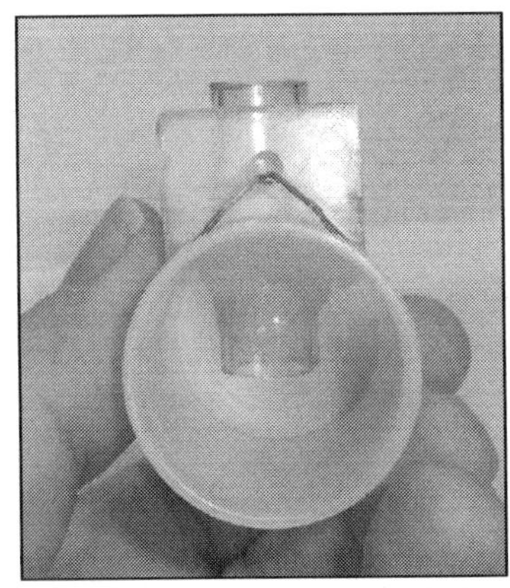

Figure 87 - Looking inside the drive cylinder.

The drive cylinder can be glued in place on top of the pressure chamber after the glue has cured. There are two ways to glue it on. If you can get a flush fit between the mounting base of the drive cylinder and the pressure chamber top you can glue it on with acrylic cement using the capillary cementing technique described earlier. If you don't have a great fit, you can glue it on with epoxy. Remember to scuff up any shiny surfaces with sandpaper before using epoxy or it will not stick to the acrylic.

The open end of the drive cylinder should be aimed slightly to the right side of the main drive axle. The pushrod will be attached about 1/4" to the right of center on the drive axle. Aim roughly at that point. The goal is to situate the drive cylinder so that it is pushing straight at its center of effort. Use masking tape to hold the drive cylinder in place until the glue is fully cured.

### *Thermoform the Drive Diaphragm*

Using a shaped diaphragm is necessary for optimal performance. One might assume this is a complicated process, but it is actually quite simple. Thermoforming tools and techniques were outlined earlier in chapter 7. Here are the steps needed to create a thermoformed diaphragm:

- Make a "vacuum table" from a flat board and a vacuum cleaner.
- Make a "form" over which we will mold the vinyl material.
- Mount the vinyl in a frame and heat it.
- Press the warm vinyl over the form and apply a vacuum.

**Figure 88 - The vacuum table is a board with holes in it that is connected to a vacuum cleaner**

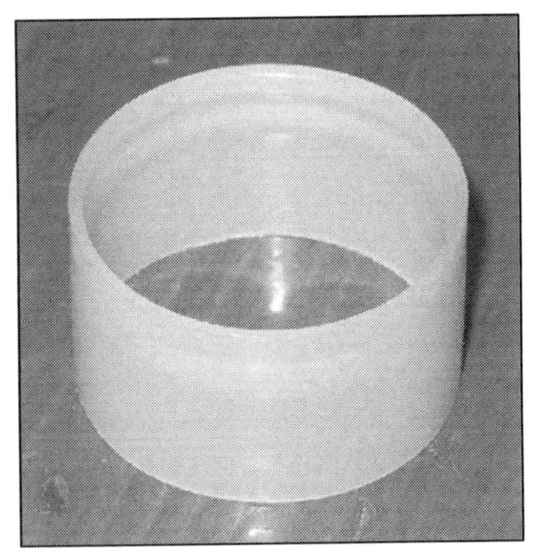

**Figure 89 - The "form" is a section cut from a 35mm film can.**

The finished diaphragm will be a cylindrical shaped dome that will fit easily inside the film can and is between 1/2" and 7/8" tall.

The "vacuum table" is not really a table at all. It is a board with some 1/8" holes drilled through it, and a vacuum cleaner hose pressed against the back side. The board needs to be situated so that it has a level surface on the top a vacuum hose attached to the bottom side. The picture in chapter 7 shows the board hanging over the side of the table, being held in place by a paint can. The vacuum cleaner hose is held against the table leg with a bungee cord. The hose is situated so that it is in contact with the bottom of the board. When the vacuum is turned on the board gets drawn up snug against the end of the hose and air is drawn in

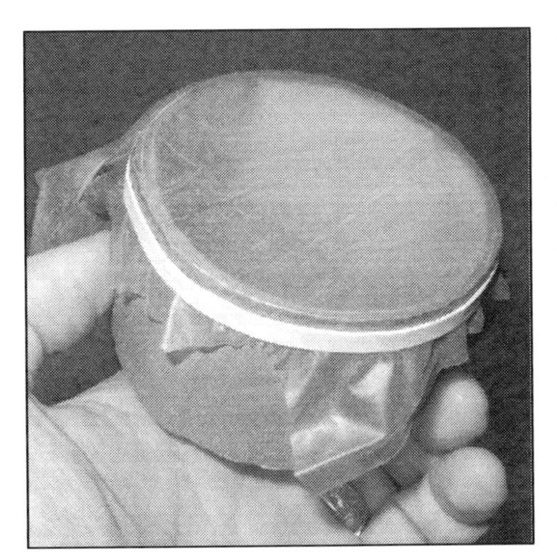

Figure 90 - Vinyl stretched over an old roll of masking tape. It looks like a little drum.

through the 1/8" holes. I usually place a paper towel over the board to create a smooth work surface.

The "form" is the object our vinyl will be molded around. The form is made from a section of plastic film can, the same kind of film can used to make the drive cylinder. Take the lid off of a 35mm film can and mark a line around the outside that is 3/4" down from the open end. Cut carefully along the line. That 3/4" tube will be used as the form for the diaphragm.

Figure 91 - The warm vinyl is drawn to the shape of the form during thermoforming.

One vinyl glove will provide enough material to make two diaphragms. One can be made from the palm of the glove, and another can be made from the back. Carefully cut the glove into two large flat sheets of vinyl. The fingers of the glove will not be used. They can be cut off and discarded.

A framework of some kind is needed to hold the vinyl flat. I use an empty roll of masking tape and a rubber band as an improvised hoop frame. Place a sheet of vinyl over the end of the old tape roll and hold it in place with a rubber band. Smooth out as many wrinkles as you can. It will look like a little drum.

If you have all those parts in order, and if you have a heat source to warm the vinyl, follow these steps to create the thermoformed diaphragm:

1. Set up the vacuum table. Place a paper towel over the holes to create a smooth work surface. Place the form in the center of the vacuum area. The tube should be placed so that the smooth "factory end" is up, and the edge that you cut is setting on the paper towel. Turn on the vacuum.
2. Stand close to your vacuum table as you heat the vinyl. Use the low setting on the heat gun. Watch the vinyl carefully as it is heated. The material will appear to be wrinkle free and relaxed when it is warm enough for forming. Back off a little with the heat gun when you notice the vinyl is wrinkle free and relaxed, but keep it warm as you move on to the next step.
3. Press the vinyl down over the form and flush to the surface of the vacuum table. The vinyl will instantly conform to the shape of the film can tube. Remove the heat. Turn off the vacuum after about 10 seconds.

Figure 92 - Cut away the excess vinyl. This thermoformed diaphragm will now be a perfect fit with the film can drive cylinder.

If you get the vinyl too hot it will melt and there will be holes in the diaphragm. Try it again with a little less heat. It may take a few tries to learn the technique. After you get one to work, make one or two more. The second and third ones are always easier, and it never hurts to keep a spare diaphragm around.

Remove the vinyl from the frame and carefully cut away the excess material. Carefully coax the thermoformed material into its new shape if necessary.

### *Create and Attach the Piston/Pushrod Assembly*

The next step is to create a lightweight plastic piston that will be attached to the diaphragm, and then create an adjustable pushrod that will connect it to one of the small holes drilled in the crank arm.

### *Piston Disks*

The piston disks are cut from a disposable coffee cup lid. If you don't have one around the house, now would be a good time to go out and get a cup of coffee! Use a quarter to trace two circles on the coffee cup lid and carefully cut them out. Scissors work well for cutting this material. Smooth off any jagged edges so that there are no sharp points to puncture the thin vinyl of the diaphragm. Carefully mark the center of both disks and drill a hole that is the same size as your attachment bolt. Lay one plastic disk on top of the other and drill the holes at the same time. This will guarantee perfect alignment.

Figure 93 - Use a quarter to trace and cut two round disks from a disposable coffee cup lid.

If you don't get the holes perfectly centered or aligned the first time, use that as an excuse to go out for another cup of coffee.

### *Attachment/Adjustment Bolt*

The attachment bolt is a 4-40x1" flat slotted machine screw with three matching nuts. The first nut holds the plastic disks together with the diaphragm material sandwiched in between them. The second and third nuts are used to attach the pushrod.

One disk will be attached to the inside of the diaphragm, and one will be attached to the outside. The attachment bolt will pass through the hole in the center of the disks and hold the entire assembly together. A small drop of silicone glue will seal the hole that is punctured in the center of the diaphragm to prevent any leaks. Here is the process for assembly:

- Place one plastic disk on the inside of the diaphragm. It should be easy to center it in the depression caused by thermoforming.
- Place a small drop of silicone glue on the outside of the diaphragm at the center, where the hole will be.
- Press the second disk onto the outside of the diaphragm and line up the holes in the two disks.
- Use a sharp object such as a needle or a piece of wire to puncture a hole in the diaphragm for the attachment bolt.
- Thread the attachment bolt through the hole from the outside so that the bolt is protruding to the inside of the diaphragm.
- Thread a nut onto the bolt until the two disks are held snuggly together and the silicone is spread sufficiently to form a seal.
- Set the assembly aside until the silicone is cured.

**Figure 94 - A thin plastic disk is bolted to the inside and outside of the diaphragm.**

### *Lower Pushrod*

The lower pushrod is fashioned from a length of 1/16" music wire. I like the music wire because it is incredibly stiff and will be perfectly straight. Some people may prefer to use a softer wire that is easier to bend, and that is OK. 1/16" craft wire will also work. It is easier to bend into shape but is not as pretty! Start with a piece of wire that is about 7" long. It will be trimmed to a shorter length when finally installed.

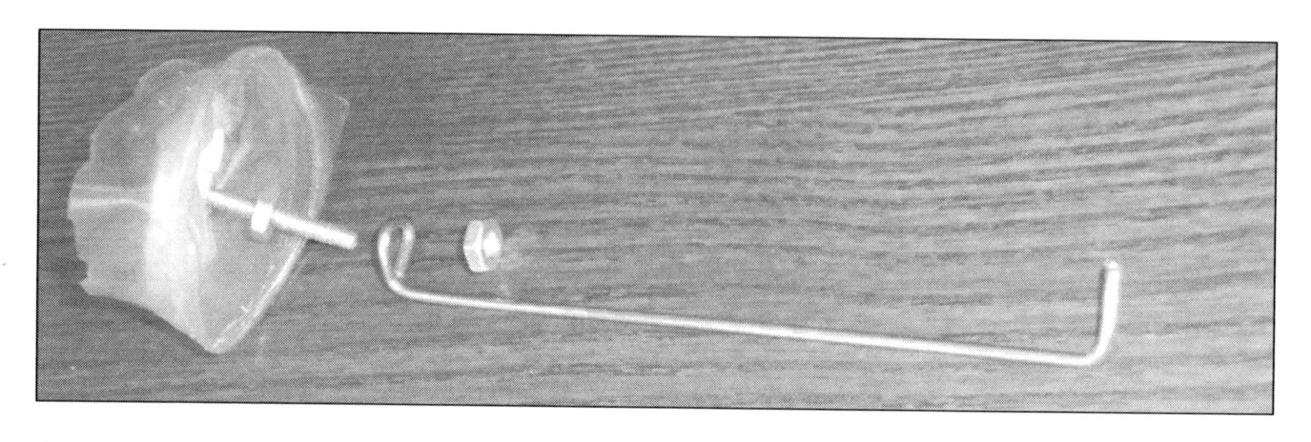

**Figure 95 - Diaphragm with attachment bolt and the lower pushrod. The eye is on the left.**

Use needle nose pliers to create an eye in one end of the pushrod wire, just big enough for the bolt to pass through. Use a heavy set of pliers to bend the eye over to 90°. This eye is used to mount the pushrod to the attachment bolt of the diaphragm/piston assembly. Thread one nut onto the attachment bolt so that it is centered on the exposed bolt. Insert the bolt through the eye of the pushrod and use a second nut to hold it snug against the first nut. This only needs to be finger-tight. If you over tighten this it will be difficult to make adjustments later.

You must now do a dry fit of the drive assembly onto your pressure chamber to determine the length of the lower pushrod. Set your motor on a level surface with the drive axle in place, and with the first crank arm attached to the drive axle. Set the crank arm level with the center of the drive cylinder and at 90° on the right side of the drive axle. Gently insert the diaphragm into the drive cylinder with the pushrod attached. Secure the diaphragm with the film can lid. We will now measure the pushrod to determine where the bend needs to be so that it will mate properly with the crank arm.

Look carefully at the pictures that show the completed assembly. Note that the crank arm is resting to the side of the drive axle. When everything is in "neutral", the crank arm is at a 90° angle from the direction of the pushrod, and the piston is midway between the ends of its normal traveling distance. The pushrod must be cut to the proper length so that the piston can travel freely *inside* the film can. The diaphragm should never be maxed out (pushed in) all the way during the normal operation of the motor. In like manner, the piston disks should never come in contact with the lid of the film can, and they should never be pulled to the outside of the film can during the normal operation of the motor.

From the neutral position you should be able to rotate the reciprocating drive axle 45° in either direction without bottoming out the diaphragm and without pulling the drive piston out past the lid.

It works best if the vents are open when you are making this measurement.

Work the mechanism back and forth to determine the best working length for the lower pushrod. Use a permanent marker to mark the spot on the pushrod where it needs to be bent. Remove the diaphragm from the engine assembly and make the final bend in the end of the pushrod. Trim off the excess pushrod so that there is about 1/4" after the bend. Reassemble the diaphragm/pushrod assembly and test it for length. If it is not exactly where it needs to be, use the nuts on the attachment bolt to adjust the length.

### *Upper Pushrod*

The upper pushrod connects the crank arm at the top of the reciprocating drive axle to the crank arm at the top of the flywheel drive axle. The crank arm on the top of the reciprocating axle is one of the two identical cranks made from 3/32" acrylic. The crank arm on the top of the flywheel axle is made from 1/4" acrylic.

Place your engine on a flat level surface and with both drive axles in place. Center both drive axles so that they are vertical and centered on their bearings. Carefully measure the distance between the top of both axle shafts. The upper pushrod needs to be that exact length.

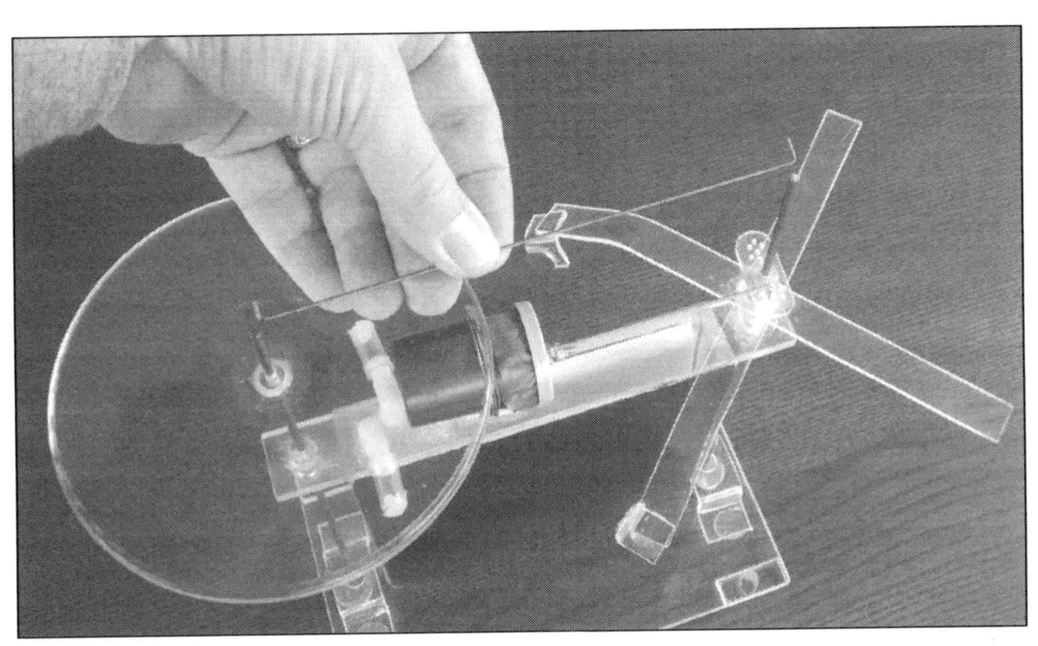

**Figure 96 - The length of the upper pushrod is exactly the same as the distance between the two axles. The tail on the flywheel end is 1/4" long. The tail on the swing arm end (rt.) is 3/8" long.**

### *Build and Attach the Flywheel*

Your flywheel doesn't have to be round. I like to make the flywheel from a round disk of clear acrylic because it looks so nice. But quite frankly it would work just as well if it was a square or a triangle as long as it was well balanced and about the same weight. If you are using simple hand tools the round acrylic flywheel is a lot of work to make. It works just as well if it is made of wood.

For those who wish to make a clear round flywheel, here are the steps I took to make it:

Figure 97 - If you have a band saw you can make a simple circle cutting jig to make a round flywheel.

- The finished flywheel should be approximately 6" in diameter, 1/4" thick, and have a 1/8" hole in the center.
- Cut a piece of 1/4" acrylic sheet to just over 6" square. Do not remove the protective coating.
- Mark the center of the disk and drill the 1/8" hole. (Use a drill bit that is approved for acrylic material)
- Make a circle cutting jig for the band saw. A circle cutting jig in this case is a large scrap of wood with a small finish nail driven into it. The board is clamped to the band saw table so that the nail is 3 inches to the right of the blade.
- Start cutting the circle free-handed until you can align the hole in the acrylic with the pin in the jig. When the acrylic is on the pin rotate the material slowly in a clockwise direction until you have cut a perfect circle.
- Remove the disk from the jig. Do not remove the protective coating. Begin sanding the edge of the disk with 100 grit sandpaper. Use several finer grades until you finally finish with 400 grit or 600 grit.
- Polish the edge with plastic polish or rubbing compound.
- Now you can remove the protective coating.
- Attach a 1/8" shaft collar to the center of the disk, aligned with the 1/8" hole that was drilled earlier.

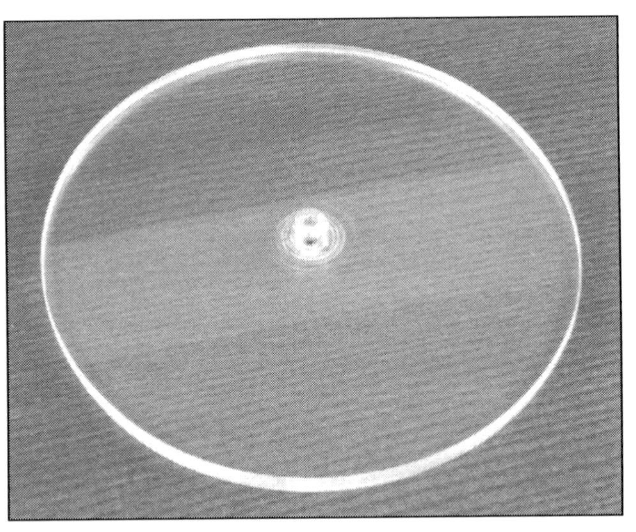
Figure 98 - 6" flywheel made from 1/4" acrylic sheet with a steel shaft collar attached at the center. There is a 1/8" hole through the flywheel that is aligned with the shaft collar.

The flywheel is the last item attached to the drive assembly. It rests on the top of the drive axle above the swing arms for the magnets. My wife tells me that it makes my engines look like the starship Enterprise.

## *Pre-Flight Checklist*

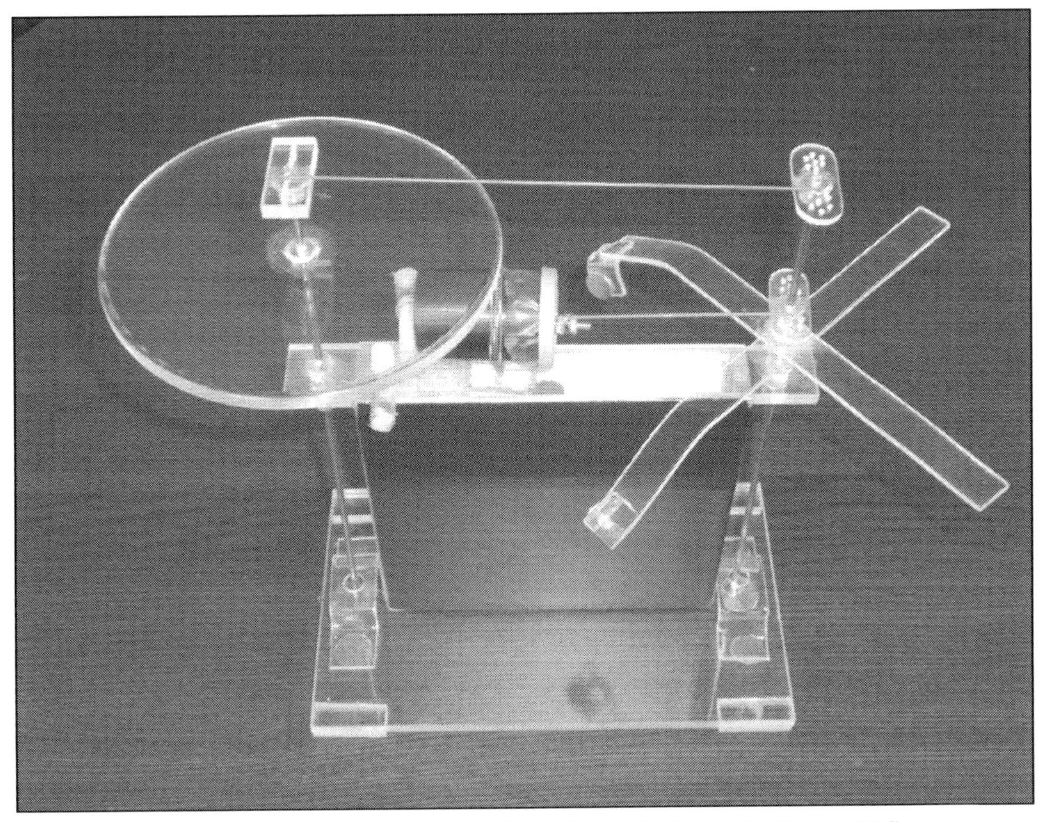

Figure 99 - Engine #2 is complete with all parts in place. You are looking at the "warm side".

Let's just double check to make sure everything is in order. It is time to re-attach any parts you have been taking on and off during assembly so that we can start your engine for the first running!

You should now have an assembled Stirling engine with the following components:

- A sealed pressure chamber mounted vertically on a stand. The aluminum plating on the sides is painted black.
- The displacer inside the pressure chamber can rock easily from one side to the other. If you swing the magnetic swing arm close to the top of the pressure chamber it will push the displacer to the opposite side.
- The reciprocating drive axle is held at the top in a glass bead bearing, and the rounded bottom end of the drive axle is resting on the smooth part of a penny. Mounted on the drive axle are the magnetic swing arms, and the crank arm.
- The flywheel is mounted on a second axle behind the drive cylinder. That axle also has a glass bead bearing at the top and a penny bearing at the bottom.
- The drive cylinder is attached to the top of the pressure chamber and the tube that connects it to the pressure chamber has an air tight seal.

- The drive assembly (diaphragm, piston, lower pushrod, and crank arm) are installed and adjusted correctly.
    - The crank arm is in its neutral position at 90° to the body of the engine and the piston pushrod.
    - The crank arm is mounted level with the center of the drive cylinder.
    - The diaphragm has an air tight seal to the film can.
    - The diaphragm needs to be set so that the piston can freely travel at least 1/8" in each direction out of the neutral position. Adjust the pushrod length to allow for proper piston movement.
- The upper pushrod connects the crank arms at the top of each axle.
- The pushrods are adjusted so that as the flywheel is rotated, the swing arms move and cause the displacer to move back and forth inside the pressure chamber.
- Close the vents when the piston is in neutral position.

## Operation and Fine Tuning

Check for pressure leaks. The easiest way to do this is to note the position of the diaphragm when sitting idle. Next, pull out on the pushrod for a few seconds and then release. The diaphragm should return to the same position it was in before the test. If it does not return to the same position there may be a pressure leak. Check all the joints and seal up any leaks if they exist.

A good starting position for the magnetic drive arms is to cross them at about 90° and make sure they are evenly distanced from the magnet in the displacer. Rotate the flywheel several times and cycle the magnetic drive assembly back and forth a few times to see that the action on each side of the displacer is about equal.

I will sometimes lubricate the drive axle at the point where it passes through the pressure chamber top plate and at the point where it touches the penny. I spray a small amount of silicone lubricant onto a Q-tip and dab a small amount on the axle at these two points.

The pushrod and diaphragm need to be adjusted so that they do not restrict the movement of the flywheel mechanism. Rotate the drive axle back and forth so that you can observe the displacer move back and forth in both directions. The drive piston needs to be able to travel freely without being restricted by the diaphragm material. If you feel resistance coming from the diaphragm (because it gets too tight, or because the piston hangs up on the cylinder) then adjust the pushrod length until there is no resistance felt.

If you have assembled your engine as illustrated, the "warm side" is the side that does *not* have the stand braces. Another indicator is that the warm side of the engine is the same side that the crank arm is on. This engine will not run if the temperature differentials are reversed. You must always apply heat to the warm side. This side of the engine is active when the displacer is on the other side of the chamber. Air expands when it is on the warm side, pushes on the diaphragm, rotates the drive axle and eventually moves the displacer to cover the warm side plate.

While it is true this engine will run from the heat of a warm hand, this is not the most rewarding thing to try first. That exercise will come later. The easy way to get your engine started is to shine a 25 to 60 watt incandescent bulb a few inches from the warm side of the engine. If you need an additional energy boost during your initial tuning, place a bag of frozen peas on the cold side of the engine. This will increase the temperature differential and help you get started with the tune-up process. After you have some time to observe and adjust your drive mechanism you will be able to run the engine without any ice.

The engine will warm up in 1 to 2 minutes and will need a slight push to get started. Rotate the flywheel in either direction to start the engine. If it does not start running right away, allow another minute or two for the engine to warm up.

It is sometimes easier if you disconnect the upper pushrod and start the engine without the flywheel. This will give you an opportunity to set and adjust the lower pushrod and the magnetic swing arms. Get the reciprocating mechanism adjusted and operational first, and then connect the flywheel and restart the engine.

I mentioned earlier I invested about $20 in a small infrared thermometer. I can point it at the surface of the engine and get an immediate temperature reading. This engine should idle along well for you the first time with a temperature differential of 30° to 40°. After you do some fine tuning it should consistently start and run with a temperature differential of 30° or less. When fine tuning is completed this engine will run with a temperature differential of 20° or less. The engine I built as a demonstration for this chapter will run down to a temperature differential of 13°. If the pressure chamber is filled with helium it will run on a temperature differential of about 10°.

I should point out that I have built several of these and I don't get the exact same performance every time. I have pointed out many times that there are variables in the building process that can have a big impact on your final results. You must build with great care to achieve a 10° to 13° temperature differential operation.

I need to also point out the way temperature differentials are measured. They are measured on the surface of the running engine, not on the source of the heat. Right now my hand is 93.3°. The room temperature is 70. But when I use my hand to run the engine in this environment I can warm up one side to about 86°, and the temperature of the other side will rise to about 75°. That is just shy of the 13° needed to make it run. In order to run my engine by the heat of my hand in here today I would need to drop the room temperature a few degrees or add helium to the pressure chamber.

Here are the adjustments you can make as you tune your engine for peak performance:

1. Adjust the length of the pushrod on the drive piston to change the relative position of the power stroke within the cylinder.

2. Open a vent to equalize the pressure after the engine has warmed up. Close the vent with the drive piston in neutral position.
3. Move both magnet arms so that the magnets are closer to the pressure chamber to shorten the throw of the engine. (Short stokes make the engine work well with smaller pressure differentials.)
4. Move both magnet arms so that the magnets are farther away from the pressure chamber to lengthen the throw of the engine. (Long throws make the engine work better with larger pressure differentials.)
5. Move one or both magnet arms in order to center them around the pressure chamber.
6. Rotate the crank arm slightly to re-center the drive arms around the pressure chamber.
7. Change the position of the lower pushrod.
8. Change the position of the upper pushrod to adjust the power ratio between the two axles.

The trick to tuning is to observe the engine carefully and make one adjustment at a time. If it improves performance, keep it. If it doesn't help, change it back. If it degrades performance, try changing it the other way. I always watch the movement of the diaphragm as I tune the engine. It is usually easy to tell what the engine is doing by watching the pressure changes on the diaphragm.

My favorite way to run these engines is to set them in the sunshine. On a 70 degree day in the Pacific Northwest the sun will heat the warm side of the engine to as much as 120°, providing a 50° temperature differential. This works better than artificial heat sources, such as an incandescent light.

### Adding Helium

There are 2 vents in case you want to try running your engine with helium. I recommend you start using your engine with air. This will help you get a feel for how to tune it and make it work. Then add helium to see the performance difference that it makes. To add helium, tip the engine so that the vents are on the bottom edge of the pressure chamber. Open both vents. Connect one vent to a helium balloon and discharge about 1/2 of its contents into the engine, then seal the pressure chamber by closing both vents. This process has now flushed most of the room air out of the engine and replaced it with the lighter helium.

When running on air or helium it is sometimes necessary to open one vent briefly so that the interior of the engine can equalize pressure with the environment. Changes in the weather bring changes in barometric pressure. Your pressure chamber is a crude barometer and is sensitive to these atmospheric pressure changes. If you are careful you can equalize the pressure several times before you notice a loss of performance caused by a loss of helium.

***If your engine won't start, take a look at some of these trouble shooting tips:***

| Symptom | Possible Cause | Solution |
|---|---|---|
| Engine will not start running | Temperature differential is too low. | Increase temperature differential by adding heat to the warm side or ice to the cool side. |
| | There may be a pressure leak. | Check all joints, seams, vents, etc. and seal any leaks. |
| | There may be too much friction. | Check all bearing points. Make sure the penny is in place. Apply minute amounts of silicone based lubricant if needed. Check the diaphragm assembly to make sure it is not hanging up on something. |
| | Heat is on the wrong side. | Heat the same side of the engine the pushrod is mounted on. |
| Displacer spends most of its time on one side, and only briefly moves to the other side. | Magnetic drive alignment problem. | Move the magnet farther away from the side that is getting only brief contact. You may need to briefly open a vent to reset the piston to its neutral position. |
| Displacer only cycles slowly a few times and then stops. | The magnets may be too close. | Move both magnets farther away from the displacer. |
| The engine starts but does not run for more than a few cycles. | The flywheel may be too light. | Tape several coins (evenly spaced for balance) to the flywheel. |
| The piston does not move far enough to make the displacer move. | Magnets are too far away or too weak. | Move magnets closer to the displacer or try stronger magnets. |
| | The crank arm setting is too long. | Shorten the crank arm if you can. |
| The flywheel reciprocates back and forth and will not continue to rotate in one direction. | Upper pushrod adjustment problem. | Lengthen the crank arm adjustment on the swing arm or shorten the crank arm adjustment on the flywheel. |
| | There may be excess pressure inside the pressure chamber. | Open a vent, move the piston to the neutral position, and close the vent. |

With a little fine tuning the engine will run on the small light bulb or sunshine with very little intervention from you.

**Running from the Heat of Your Hand**

Remember that the engine does not run on heat alone. It runs on a temperature differential. There must be heat, and there must be something cooler. If you are in a very warm room, or if it is a hot summer day, the environment will not be cooler than the palm of your hand. It is also true that not all people have warm hands.

It is easiest to start and run one of these engines from the heat of your hand when the engine is cool, and when it is in a cool environment (below 70°). Place the engine on a level surface. Rest the palm of your hand against the warm side of the engine. Pay close attention to the drive mechanism and you will see it move slightly as the engine warms up. After a minute or so, open one of the vents and re-center the drive mechanism, then close the vent.

Give the engine a little push to get it started. If it doesn't start right away you may need to continue warming for a little longer. If your hand starts to get cool, turn the engine around so you can place your other hand on the warm side and continue warming the engine. It should start with a little encouragement from you in about another minute.

The magnet drive arms usually have to be situated a little closer to the pressure chamber when running from the heat of your hand. After a little practice you will learn how to set up your engine for running with a variety of different temperature differentials.

This process is much easier when helium has been added to the pressure chamber. An engine filled with helium will run in a warmer room because it does not require as great of a pressure differential.

A helium charge seems to last from 7 to 10 days. After that time the performance seems to return to the room-air baseline.

# Chapter 11: Engine #3 - Vertical Flywheel Magnetic Drive Stirling Engine

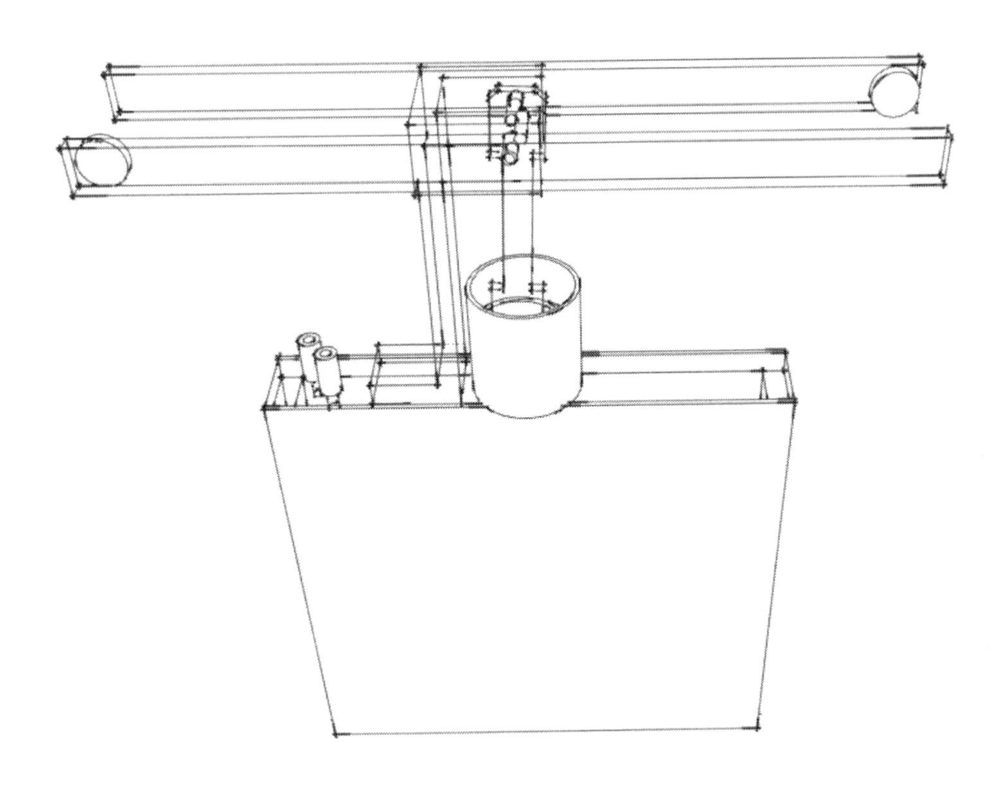

## Engine Design Explained

Engine #3 is an attempt to make a Stirling engine with a rotating flywheel and with a minimum number of moving parts. Unfortunately this engine will probably not run from the heat of your hand, but it does very well running in the sun, in front of a light bulb, or with ice.

This engine is different from the first two in that the flywheel is oriented vertically. This vertical orientation causes more drag on our primitive bearings which has the potential for a slight decrease in performance. For this reason the engine has a lighter flywheel.

The flywheel is actually two arms that are joined by a single axle that also has a crankshaft. There is a magnet mounted on each arm that moves the displacer as each magnet passes the bottom of its rotation cycle. The magnets are positioned 180° apart, providing almost 90° of propulsion for both the push and pull stroke of the drive diaphragm. Every time a magnet on one of the arms passes by the pressure chamber it causes the displacer to change sides. When the displacer forces the air inside the chamber to the warm side of the engine it expands and pushes on the drive cylinder and rotates the crankshaft. When the displacer moves the air to the cool side it causes the pressure inside the chamber to drop and it pulls on the crankshaft, completing its rotation.

Mounting the drive axle to a horizontal axis permits the drive cylinder to take on a vertical orientation. This allows for a simpler mounting strategy and also prevents gravity from pulling the drive mechanism off center as it can do in the first two designs.

## Parts List

The table below represents the retail prices in March of 2010. Many of the supplies I purchased came in large quantities that will provide enough material to make several engines. Because of that, I have listed 2 prices. The first column is the price I paid for the material. The second column represents the adjusted cost for the amount of materials that were actually used in this engine.

| **Engine #3** | | | |
|---|---|---|---|
| **Item** | **Count** | **Retail** | **Actual** |
| **6" x 12" x 0.064 Aluminum Sheet** | 1 | $ 7.29 | $ 7.29 |
| **1/8" (0.125) Music Wire** | 1 | $ 1.16 | $ 0.19 |
| **3/32" (0.093) Clear Acrylic** | 1 | $ 1.88 | $ 0.19 |
| **1/4" (0.220) Clear Acrylic** | 1 | $ 14.97 | $ 3.74 |
| **1/4" Nylon Barb Splicer** | 1 | $ 0.89 | $ 0.89 |
| **Vinyl Gloves (10 pk)** | 1 | $ 3.97 | $ 0.40 |
| **1/8" Steel Shaft Collar** | 6 | $ 1.00 | $ 6.00 |
| **1/4" ID Vinyl Tubing (10')** | 1 | $ 3.11 | $ 0.05 |
| **35mm Film Can** | 2 | $ - | $ - |
| **Neodymium Magnets 1/16" x 1/2"** | 4 | $ 0.45 | $ 1.80 |
| **JB Weld Epoxy** | 1 | $ 4.97 | $ 0.99 |
| **5 Minute Epoxy** | 1 | $ 3.97 | $ 0.79 |
| **Clear Silicone** | 1 | $ 3.89 | $ 0.97 |
| **3/16" Foam Poster Board** | 1 | $ 3.62 | $ 0.08 |
| **Black Spray Paint** | 1 | $ 0.97 | $ 0.24 |
| **Helium Filled Balloon** | 1 | $ - | $ - |
| **4-40 x 1 Machine Screws and Nuts (10pk)** | 1 | $ 0.99 | $ 0.33 |
| **Acrylic Cement** | 1 | $ 5.30 | $ 0.53 |
| **1/8" ID Glass Bead** | 3 | $ 0.10 | $ 0.30 |
| | | $ 58.53 | $ 24.79 |

Since I am a bargain hunter, I will always shop around for the best prices. If you are lucky enough to find scrap acrylic for $1.00 a pound and a good discount hardware store you should be able to bring that $58.53 figure down to about $38.00.

## Conceptual Drawings

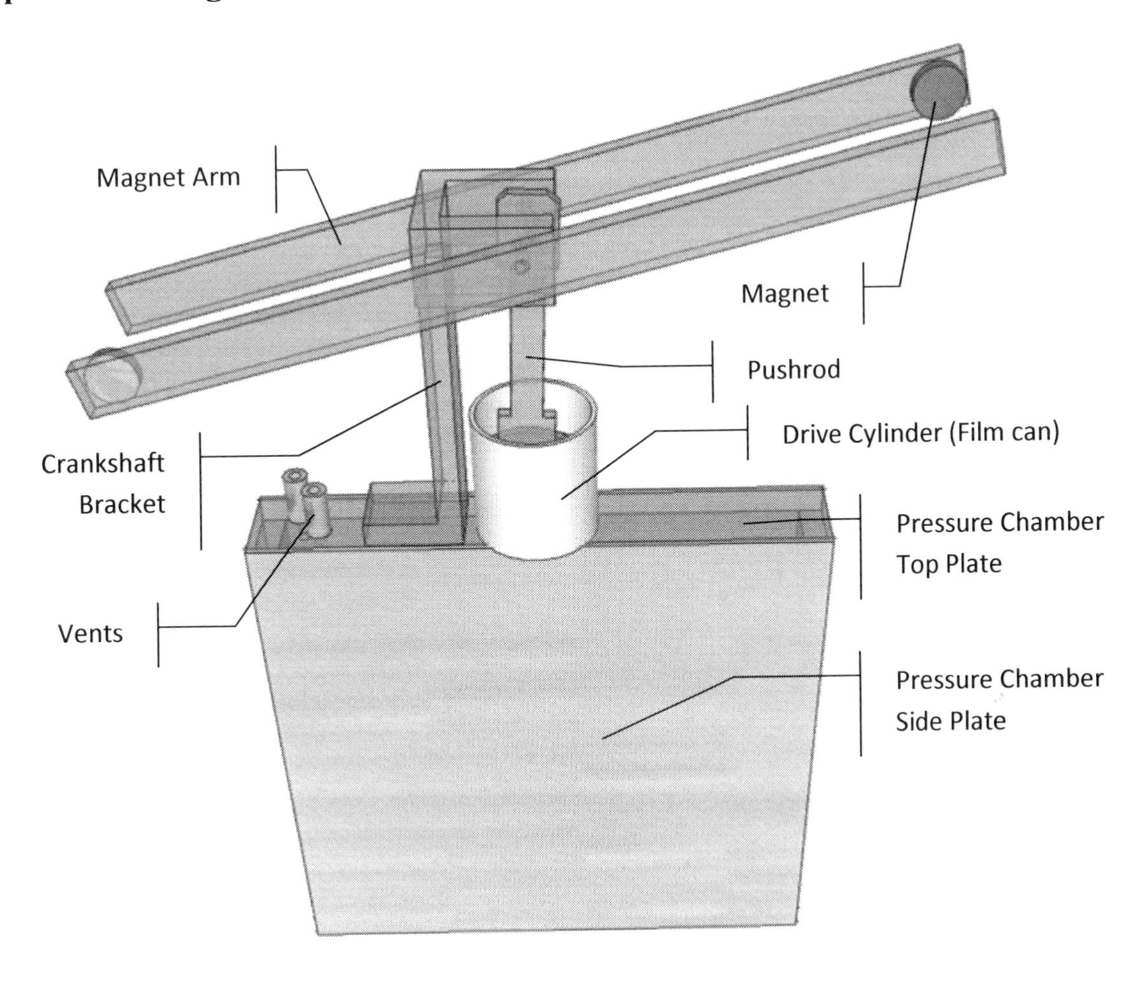


I will often create a conceptual model as part of my Stirling engine design process. This drawing was made using a free program called Google SketchUp. It is a great way to visualize a project before you start to build it. The finished engine has a few things on it that are different from this picture. You will notice the crankshaft bracket and the pushrod have evolved a bit from the original concept. The stand is not included in this drawing as it was not a functional part of the original concept drawings. This view does not show you the drive piston/diaphragm assembly or the crankshaft.

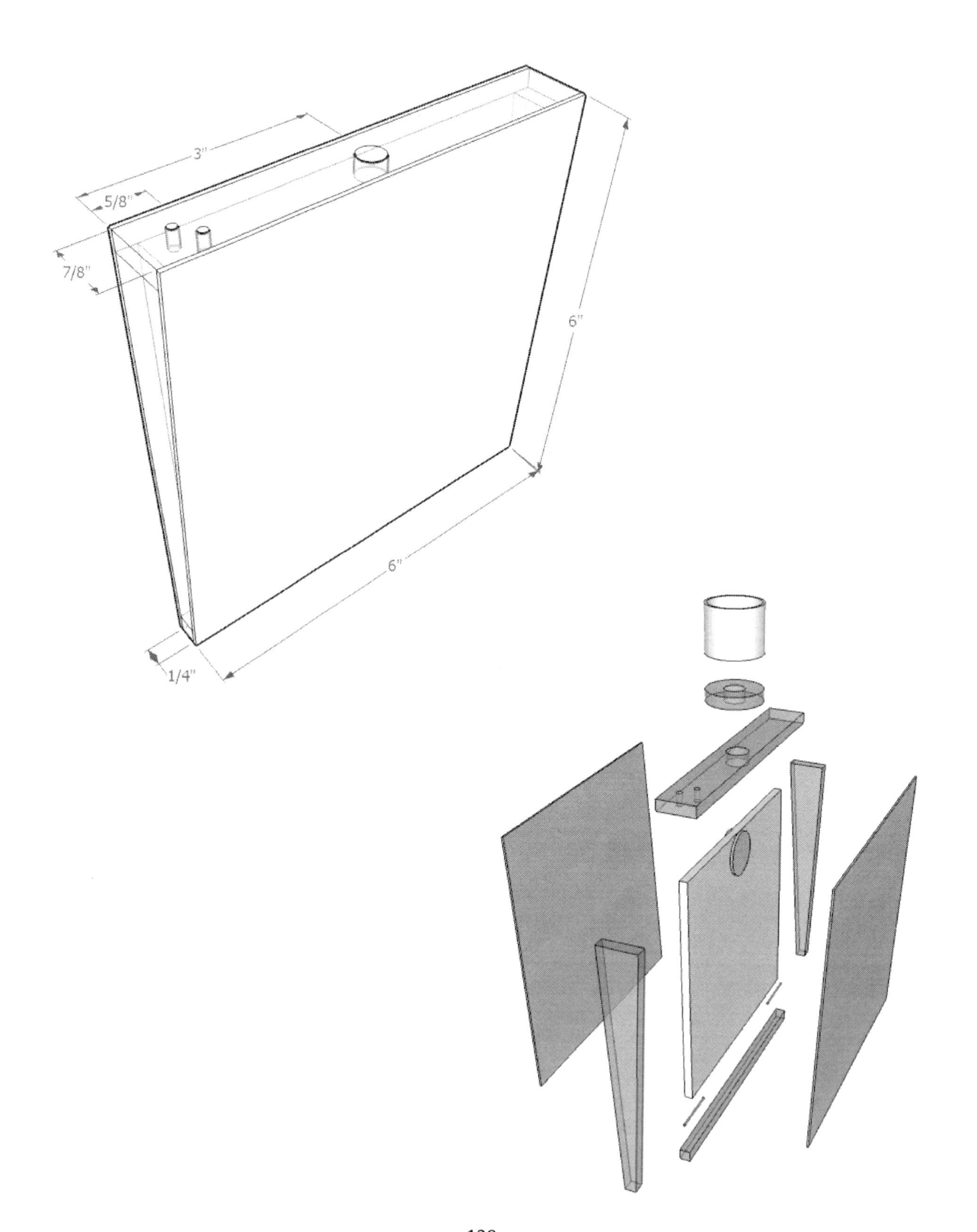
3"
5/8"
7/8"
6"
6"
1/4"

## Step by Step Instructions

### *Draw Plans and Create Templates*

Here is a detailed drawing of the pressure chamber parts and their measurements. You will find it very helpful if you take some time and draw your own full scale plans and templates before you begin. The templates are particularly useful for creating the acrylic parts of the pressure chamber. I generally begin each engine project with some full scale drawings of the parts I need to make. The vent holes are 1/8". The hole for the drive cylinder is 1/2". There is no round flywheel on this engine.

**Figure 100 - The templates for two pressure chambers are laid out on the acrylic sheet in preparation for cutting. Making full sized templates is very useful in this process.**

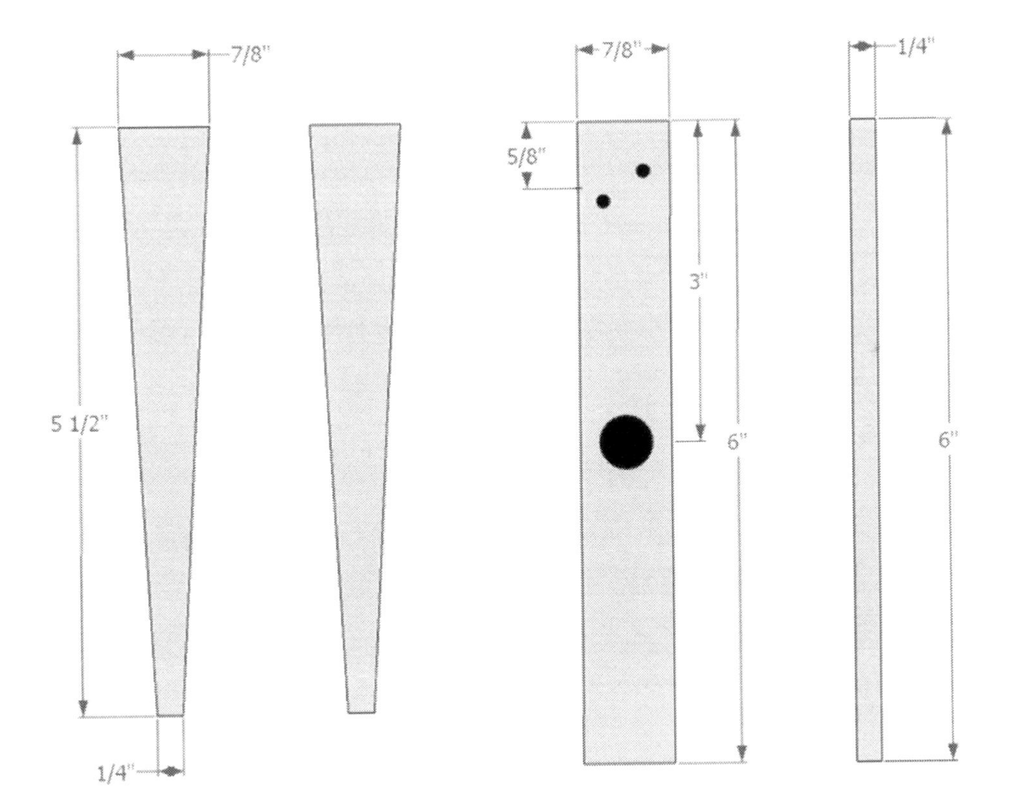

### *Cut and Paint the Aluminum for the Pressure Chamber Side Walls*

You will need to cut two 6" squares of aluminum sheet. The aluminum I purchased for this project comes in a sheet that measures 6" by 12". Only one simple cut was required.

The aluminum must be of a heavy enough gauge so that it will not easily flex. Do not use thin flimsy metal as it will not perform as well. The material here has a thickness of 0.064. Aluminum that has a thickness near 0.04" will also work. If it is much thinner than that it becomes too flexible for a pressure chamber.

**Figure 101 - Pressure chamber side walls are painted black on both sides except for the inside border (shown here).**

I don't recommend the use of tin snips for cutting flat sheets of metal. The tin snips will bend the metal on one side of the cut.

You can cut this by hand with a hack saw if you are patient, but it will be a challenge because a hack saw is not really intended for cutting material that is 6" wide.

I recommend either cutting it with a nibbler or with a metal cutting blade in a saber saw. If you use the nibbler you will need to flatten the edge of the metal after it has been cut, but that is easy to do. If you use a saber saw you will want to use a board and some clamps to hold the metal securely to your workbench and then use the board as a cutting guide. Always secure metal parts with clamps before attempting a cut with a power tool. You don't want these sharp objects flying through the air at high speed! It can get quite noisy, so hearing protection is recommended.

Please refer to the chapter, "Working with Aluminum" for a detailed explanation of cutting and finishing techniques.

Use masking tape to prevent painting the area around the inside edge of each sheet where the acrylic parts will be attached to the aluminum. This will help provide a strong glue joint. Glue sticks better to bare aluminum than it does to painted aluminum. Paint both sides of the aluminum sheets black.

The black paint really does make a difference in engine performance. I have made Stirling engines with black paint, white paint, and no paint. I now paint all sides of my aluminum sidewalls with flat black spray paint.

### *Cut the Acrylic Parts for the Pressure Chamber Bottom, Ends, and Top*

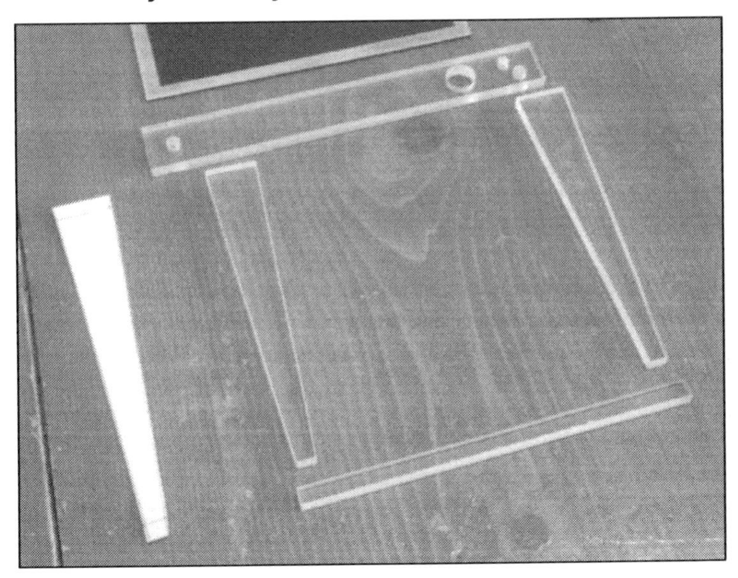

**Figure 102 - Pressure chamber top, bottom, and end pieces cut from 1/4 inch acrylic sheet.**

The pressure chamber dimensions are made to accommodate the material chosen for the displacer. My displacer material is about 3/16" thick. If you are able to locate material for your displacer that is 3/16" thick then just use the dimensions provided in the drawings at the beginning of the chapter. If for some reason you cannot find an exact match for the displacer material you must adjust the dimensions of your pressure chamber accordingly. i.e., If your displacer is 1/4" thick then you should make the pressure chamber 1/16" wider at both the top and the bottom.

Lay out the full scale pattern for each of the acrylic parts and attach it to the surface of the acrylic for cutting. Do not remove the protective coating from the acrylic sheet until after the parts are cut. Leave enough space between the parts to accommodate the width of the cut that will be caused by your saw. I used a saber saw for cutting the parts you see in these illustrations.

Use a sanding block if necessary to make all cut surfaces in the acrylic flat and straight. Both end pieces (the tapered ones) must be identical. Tape or clamp these pieces together when sanding the edges in order to maintain a consistent shape. Only sand the cut edge. Do not sand the shiny flat surface of the acrylic. If cutting leaves a buildup of melted acrylic you can sometimes scrape it off. Use another piece of acrylic as a scraper to remove the rough particles that may be stuck to the cut edge.

The long edges of the top plate and the bottom plate should be angled to match the angle of the end pieces. If you don't have the ability to cut this angle reliably, cut them square and then sand them to the correct angle. It is possible to sand the edges after the frame is cemented together by rubbing the completed frame on a sheet of flat sandpaper.

Drill the holes in the pressure chamber top plate prior to assembly. The holes are less likely to cause a crack or a break in the acrylic if you drill them before cutting. The holes for the bearings are 3/16". The vent holes are 1/8". The hole for the drive cylinder is 1/2"

### *Assemble the Pressure Chamber – Leaving One Side Off*

You may begin to assemble the pressure chamber when the paint is cured on the aluminum parts and the acrylic framework pieces have been cut to size.

**Figure 103 - Hold the acrylic pieces so that nothing touches the joint as you join the parts.**

Dry assemble all parts first to guarantee a good fit. Shape or replace any parts that are not fitting well.

Assemble the acrylic frame using the capillary cementing technique described in chapter 6. This provides an attractive and sturdy joint for the acrylic frame. When the frame is complete and the joints have cured you will attach it to one of the aluminum side plates. (The other aluminum plate will be glued on after the displacer is assembled and placed inside.)

I learned the hard way that you don't want any clamps or braces (or even tape) to be touching the joint when joining acrylic with solvent. The solvent will run under the tape or the clamp and disfigure your beautiful acrylic surface. The picture here shows how I held the parts square while cementing them together so that nothing would be touching the joints when the solvent was applied.

I recommend using a slow cure epoxy like JB Weld for attaching the aluminum side plates to the acrylic framework. The slow cure will provide you with plenty of time to work with the joint. You won't be rushed by a five minute time limit like you would with faster curing glue. Also, JB Weld is very thick and sticky, which is helpful for this operation. You want to make sure you don't get too much glue oozing out of the joint. Large globs of glue on the inside of the pressure chamber can interfere with the operation of the displacer after the engine is assembled.

**Figure 104 – Engine #3 pressure chamber frame assembly is glued to one aluminum side plate with epoxy. One side is left off to enable installation of the displacer.**

You will not be able to wipe the epoxy off the acrylic if you make a mistake. If you dribble a little glue on your acrylic, and it is not in a place that will interfere with the operation of moving parts, my advice is you leave it there. Trying to clean it off will probably just make it look worse.

### *Build the Displacer*

The Displacer will be cut from 3/16" thick poster board. Two thin neodymium magnets will be attached to the center of the top edge, and two small wires will be attached at the bottom. The wires will extend slightly past the end of the displacer and will prevent the displacer from getting close enough to touch the acrylic end plate. The wire also provides a smaller footprint for the displacer and reduces friction that would be caused by having the entire edge of the displacer on the bottom of the pressure chamber.

The dimensions of the displacer will vary slightly depending upon the thickness of the acrylic used to make the end plates. The displacer should be cut so that there is about 1/8" clearance on all sides when it is inside the pressure chamber.

**Figure 105 - Small wires are attached to the bottom edge of the displacer. The tape is removed after the wires are glued in place.**

Measure the inside height and width of the pressure chamber. Cut the poster board to be 1/4" shorter than the inside of the pressure chamber (this will make 1/8" clearance on each side).

Check the fit and mark which side of the displacer will be at the top. Attach a thin (1/16") neodymium magnet to each side of the displacer, centered along the top edge. My magnets were 1/2" round and 1/16" thick. Use silicone or epoxy to glue the magnet to the outside top edge (centered). The magnets must be installed so that they are attracting each other. This is very convenient as it makes it possible to glue them to the displacer without a clamp. The direction of the poles does not matter as long as the magnets are set to attract.

I have also built displacers with a single magnet imbedded inside the foam core of the displacer material. You can do it this way if you can't find thin magnets. You should understand that the magnets serve a second purpose when mounted to the outside edge of the displacer. They prevent the displacer material from resting flush against the aluminum sidewall. If you choose to place your magnet inside the foam core you will have to add a small spacer (such as a drop of glue) to the top edge of the displacer so that it will always stand off from the side wall about 1/16".

Cut two pieces of thin wire to a length of 1". Use sandpaper or a file to remove any burrs from the cutting process. Use tape to temporarily attach one to each bottom corner of the displacer. They should be taped to the bottom edge, so that they rest flat on the bottom of the pressure chamber when assembled.

The ends of the wire should protrude 1/16" on each side. This will act as a "stop" to prevent the displacer from getting too far out of center.

Gently place the displacer assembly inside the pressure chamber to inspect the fit. If both of the short wires are touching the sides of the frame at the same time they are too long. The displacer must be able to rock back and forth with no obstruction. Set the side plate in place and check to make sure the displacer clears all sides by about 1/8" as it moves back and forth.

When you are happy with the fit, use epoxy to glue the wires in place on the bottom edge of the displacer. Use tape to hold the wires in place and set it aside until the glue is completely cured, and then remove the tape.

**Figure 106 – The displacer is in position and is checked for size. The magnet is at the top (wide side) and the two wires are at the base (narrow side).**

### *Attach the Second Pressure Chamber Side Plate*

This is one of the more complicated and tricky parts of this assembly process. The displacer must be placed inside the pressure chamber so that the magnets are at the top (the wide end) and the wires are at the bottom (narrow end). The side plate must be glued in place to make an air tight seal, but still allow the displacer to move freely from side to side.

If glue oozes from the joint and sticks to the displacer, and the displacer can't move freely, the engine will not run.

**Here is how I do it:**

- Place the partially assembled pressure chamber on a flat surface with the open side facing up.
- Place the displacer inside the pressure chamber.
- Place the aluminum side plate in place (with no glue yet) and inspect the fit. It should be a flush tight joint.
- Remove the side plate and apply a thin coat of JB Weld 60 minute epoxy to the acrylic frame side of the joint.
- Press the aluminum side plate against the glue and apply enough pressure to squeeze the excess glue from the joint.
- Immediately remove the aluminum side plate and inspect the glue joint to make sure you have an air tight seal. You should see a continuous trail of wet glue all the way around the aluminum side plate. If there is excess glue oozing into the interior of the pressure chamber remove it now. Be careful you do not accidentally glue the displacer to the inside of the pressure chamber.

- Replace the side plate and gently clamp it with a rubber band or weighted object until the glue has cured.
- Keep the pressure chamber flat until the glue is cured. Do not pick it up or try to make the displacer move while the glue is still sticky. If you cause your displacer to move while there is still tacky glue inside it could glue your displacer to the inside of your engine and then your engine will not work.

### *Build a Stand for the Pressure Chamber*

At this point you should have an assembled pressure chamber that has a clear acrylic frame (top, bottom, and ends), two black aluminum side plates, and a displacer. The displacer should move freely from side to side, and you should be able to make it move by holding another magnet near one of the magnets on the displacer. It is now time to build some type of apparatus that will hold it upright and will hold the bottom bearing for the drive axle.

Figure 107 - Build a stand with angled vertical braces that will hold the pressure chamber with its centerline perfectly vertical. Tape is being used to hold the pieces in place for cementing.

The base can be built of wood or acrylic. Acrylic is only slightly harder to work with and makes a very attractive stand. I have used both materials and favor the acrylic for its good looks. The stand needs to provide a wide base for the assembled engine so that it does not fall over. This stand does not need to be any longer than the base of the pressure chamber. For this engine I used a piece of 1/4" acrylic sheet that measures 5 1/4" by 6 1/2". I cemented four small scraps of acrylic to the bottom of the base to act as feet. These are intended to prevent the bottom of the acrylic base from getting scratched.

The two vertical bars are approximately 1 1/2" by 3 1/2". One of the long sides on each vertical brace is cut at an angle so that the pressure chamber will stand with its centerline perfectly straight up and down. The vertical braces must be set as close to the edge of the pressure chamber as possible. Set them so that the outside edge of the brace is flush with the end of the pressure chamber.

You can find the angle for the vertical brace by tracing the same template you used to cut the pressure chamber end plates. You can use a carpenter's bevel tool to check your angles prior to final assembly. If the centerline of the pressure chamber is vertical, the angle will be the same on both sides.

### *Attach the Pressure Chamber to the Base*

Use clear silicone to attach the pressure chamber to the base. Apply a small bead of silicone adhesive along all the contact area of the base and the pressure chamber. Check the angles on both sides of the pressure chamber and when they are correct, use tape or a rubber band to hold the pressure chamber in place until the silicone has set.

Silicone is used because it is not necessarily permanent. If you ever need to disassemble this joint to correct the angle you will be glad you used silicone.

**Note:** The centerline of the pressure chamber must be perfectly vertical! Both sides of the pressure chamber must slope at the same angle. This is critical to the balance of the engine. If these angles do not match, the displacer will not move easily in both directions. Have I said this before?

### *Create and Attach Vents*

Venting is important for two reasons. It is occasionally necessary to equalize the pressure inside your motor with the pressure of the environment. If you have done well in creating a sealed pressure chamber, the diaphragm and the attached drive assembly will become very sensitive to small changes in pressure.

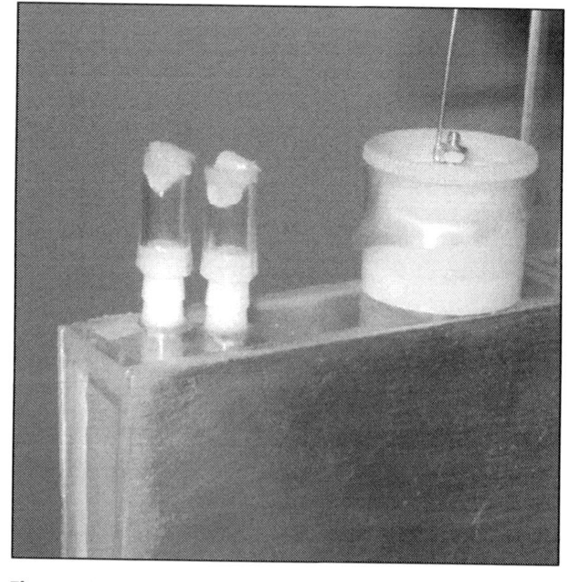

**Figure 108 - Vents are fashioned from a 1/4" nylon barb splice. One barb splice fitting can be cut to make two nipples.**

The second purpose of the vents is for those who wish to increase the performance of their engine by running it with helium inside. A carefully constructed and well tuned engine of this design will run from the heat of a warm hand in a cool room with just air inside. Adding helium makes a noticeable difference in those situations where the engine does not appear to have enough temperature differential to start running.

The vents are crafted from small fittings made for connecting 1/4" plastic tubing. They can be brass or plastic. Cut the fittings so that only the pipe nipple section remains (See figure 24). Use epoxy to glue the nipples over the two small holes on the top plate of the pressure chamber, behind the drive cylinder. As you can see in the illustration here, I sometimes use straight nipples and sometimes use ones that are angled. Remember to use sandpaper to roughen the surface of the acrylic before gluing. Do not let the glue run into the vent hole as this could cause the vent to plug, or it could drip inside and foul up the displacer. Use epoxy to glue the vent nipples to the pressure chamber top plate. Remember to rough up the surface of the acrylic lightly with sandpaper.

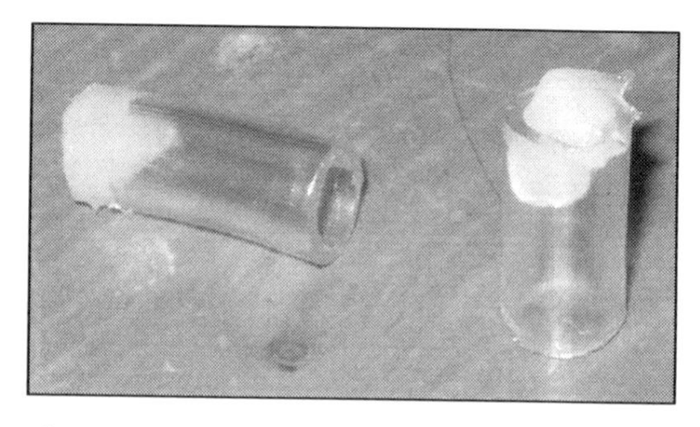

**Figure 109 - Vent plugs made from 1/4" tubing and silicone glue.**

The vent plugs are made from 1" pieces of 1/4" tubing that has had one end filled with silicone glue. Cut two of these and carefully plug one end with silicone. Set them aside until the silicone has cured.

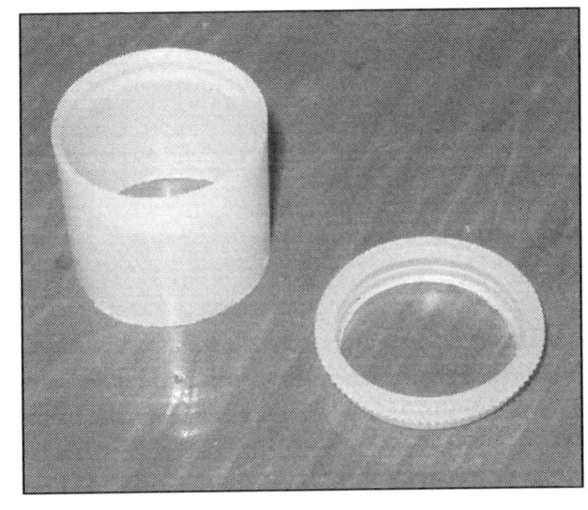

Figure 110 - Remove the center from the film can lid. Cut the bottom out of the film can so that the remaining tube is 1" tall.

### *Build the Film Can Drive Cylinder and Attach*

The drive cylinder is fashioned from a 35mm film can and lid. The bottom of the film can is removed and the remaining tube is fitted to a piece of acrylic mounted to the top of the pressure chamber.

Make a large hole in the lid of the film can. Remove the material from the center of the lid so that only a circular rim remains. There needs to be only enough material in the lid so that it will still snap back into the canister. This rim will be used to anchor the diaphragm material and hold an air tight seal.

Figure 111 - Create a disk of 1/4" acrylic that will fit inside the bottom of the film can.

Use the body of the film can to trace a circle onto a piece of 1/4" acrylic material. Drill a 1/2" hole in the center of the circle, and carefully cut out the part along the line you traced. It should easily fit in the bottom of the film can body.

Figure 112 - Attach the disk to the top of the pressure chamber. The hole in the disk must be aligned with the hole in the pressure chamber.

Attach the acrylic disk to the center of the top plate of your pressure chamber. Line up the hole in the acrylic disk with the hole in the top plate of the pressure chamber. You can cement this in place using the capillary cementing technique described earlier, or with epoxy or silicone. Allow this joint to cure before proceeding. This must be an air tight joint.

Use silicone to attach the film can body to the disk that is now secured to the top of the pressure chamber. Take care to create an air tight seal around the base of the film can. Set this aside until the silicone has cured.

### *Thermoform the Drive Diaphragm*

Using a shaped diaphragm is necessary for optimal performance. One might assume this is a complicated process, but it is actually quite simple. Thermoforming tools and techniques were outlined earlier in chapter 7. Here are the steps needed to create a thermoformed diaphragm:

- Make a "vacuum table" from a flat board and a vacuum cleaner.
- Make a "form" over which we will mold the vinyl material.
- Mount the vinyl in a frame and heat it.
- Press the warm vinyl over the form and apply a vacuum.

**Figure 113 - The vacuum table is a board with holes in it that is connected to a vacuum cleaner**

**Figure 114 - The "form" is a section cut from a 35mm film can.**

The finished diaphragm will be a cylindrical shaped dome that will fit easily inside the film can and is between 1/2" and 7/8" tall.

The "vacuum table" is not really a table at all. It is a board with some 1/8" holes drilled through it, and a vacuum cleaner hose pressed against the back side. The board needs to be situated so that it has a level surface on the top a vacuum hose attached to the bottom side. The picture in chapter 7 shows the board hanging over the side of the table, being held in place by a paint can. The vacuum cleaner hose is held against the table leg with a bungee cord. The hose is situated so that it is in contact with the bottom of the board. When the vacuum is turned on the board gets drawn up snug against the end of the hose and air is drawn in through the 1/8" holes. I usually place a paper towel over the board to create a smooth work surface.

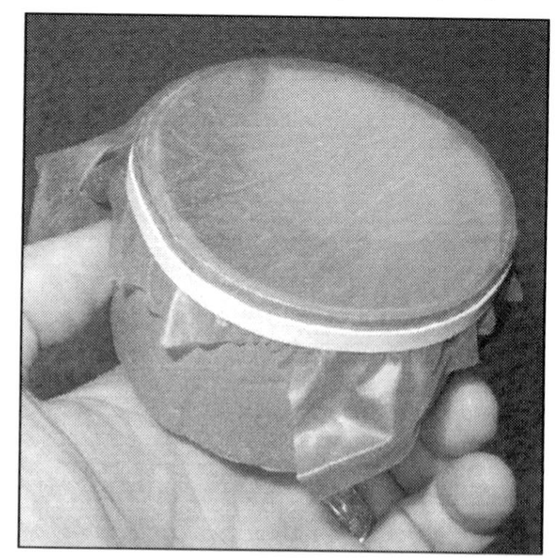

**Figure 115 - Vinyl stretched over an old roll of masking tape. It looks like a little drum.**

The "form" is the object our vinyl will be molded around. The form is made from a section of plastic film can, the same kind of film can used to make the drive cylinder. Take the lid off of a 35mm film can and mark a line around the outside that is 3/4" down from the open end. Cut carefully along the line. That 3/4" tube will be used as the form for the diaphragm.

Figure 116 - The warm vinyl is drawn to the shape of the form during thermoforming.

One vinyl glove will provide enough material to make two diaphragms. One can be made from the palm of the glove, and another can be made from the back. Carefully cut the glove into two large flat sheets of vinyl. The fingers of the glove will not be used. They can be cut off and discarded.

A framework of some kind is needed to hold the vinyl flat. I use an empty roll of masking tape and a rubber band as an improvised hoop frame. Place a sheet of vinyl over the end of the old tape roll and hold it in place with a rubber band. Smooth out as many wrinkles as you can. It will look like a little drum.

If you have all those parts in order, and if you have a heat source to warm the vinyl, follow these steps to create the thermoformed diaphragm:

1. Set up the vacuum table. Place a paper towel over the holes to create a smooth work surface. Place the form in the center of the vacuum area. The tube should be placed so that the smooth "factory end" is up, and the edge you cut is setting on the paper towel. Turn on the vacuum.
2. Stand close to your vacuum table as you heat the vinyl. Use the low setting on the heat gun. Watch the vinyl carefully as it is heated. The material will appear to be wrinkle free and relaxed when it is warm enough for forming. Back off a little with the heat gun when you notice the vinyl is wrinkle free and relaxed, but keep it warm as you move on to the next step.
3. Press the vinyl down over the form and flush to the surface of the vacuum table. The vinyl will instantly conform to the shape of the film can tube. Remove the heat. Turn off the vacuum after about 10 seconds.

Figure 117 - Cut away the excess vinyl.

If you get the vinyl too hot it will melt and there will be holes in the diaphragm. Try it again with a little less heat. It may take a few tries to learn the technique. After you get one to work, make one or two more. The second and third ones are always easier, and it never hurts to keep a spare diaphragm around.

Remove the vinyl from the frame and carefully cut away the excess material. Carefully coax the thermoformed material into its new shape if necessary.

### *Create and Attach the Piston/Pushrod Assembly*

The next step is to create a lightweight plastic piston that we can attach to the diaphragm, and then create an adjustable pushrod that will connect it to one of the small holes we drilled in the crank arm.

### *Make Piston Disks*

The piston disks are cut from a disposable coffee cup lid. If you don't have one around the house, now would be a good time to go out and get a cup of coffee! Use a quarter to trace two circles on the coffee cup lid and carefully cut them out. Scissors work well for cutting this material. Smooth off any jagged edges so that there are no sharp points to puncture the thin vinyl of the diaphragm. Carefully mark the center of both disks and drill a hole that is the same size as your attachment bolt. You can lay one plastic disk on top of the other and drill the holes at the same time. That will guarantee perfect alignment.

**Figure 118 - Use a quarter to trace and cut two round disks from a disposable coffee cup lid.**

If you don't get the holes perfectly centered or aligned the first time, use that as an excuse to go out for another cup of coffee.

### *Attachment/Adjustment Bolt*

The attachment bolt is a 4-40x1" flat slotted machine screw with three matching nuts. The first nut holds the plastic disks together with the diaphragm material sandwiched in between them. The second and third nuts are used to attach the pushrod.

One disk will be attached to the inside of the diaphragm, and one will be attached to the outside. The attachment bolt will pass through the hole in the center of the disks and hold the entire assembly together. A small drop of silicone glue will seal the hole that is punctured in the center of the diaphragm to prevent any leaks. Here is the process for assembly:

- Place one plastic disk on the inside of the diaphragm. It should be easy to center it in the depression caused by thermoforming.
- Place a small drop of silicone glue on the outside of the diaphragm at the center, where the hole will be.
- Press the second disk onto the outside of the diaphragm and line up the holes in the two disks.
- Use a sharp object such as a needle or a piece of wire to puncture a hole in the diaphragm for the attachment bolt.
- Thread the attachment bolt through the hole from the outside so that the bolt is protruding to the inside of the diaphragm.

- Thread a nut onto the bolt until the two disks are held snuggly together and the silicone is spread sufficiently to form a seal.
- Set the assembly aside until the silicone is cured.

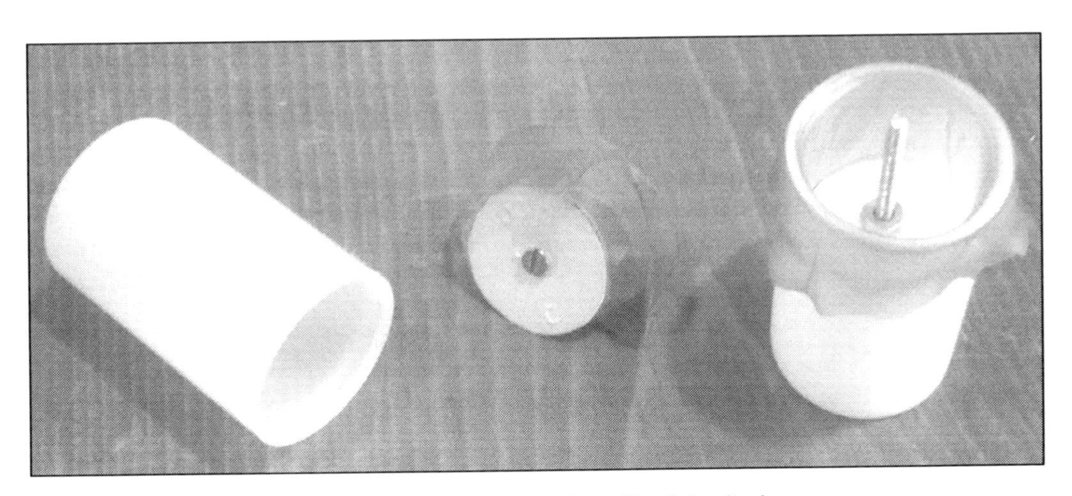

Figure 119 - A thin plastic disk is bolted to the inside and outside of the diaphragm.

### *Build the Crankshaft Bracket*

You need to build a bracket that will support your crankshaft directly above the drive cylinder at the center of your engine. I used 3/16" acrylic sheet for this because I had some lying around, but you don't need to go out and buy a separate sheet just to match the thickness. You can make this part out of the same 1/4" material you are using for the rest of the acrylic parts.

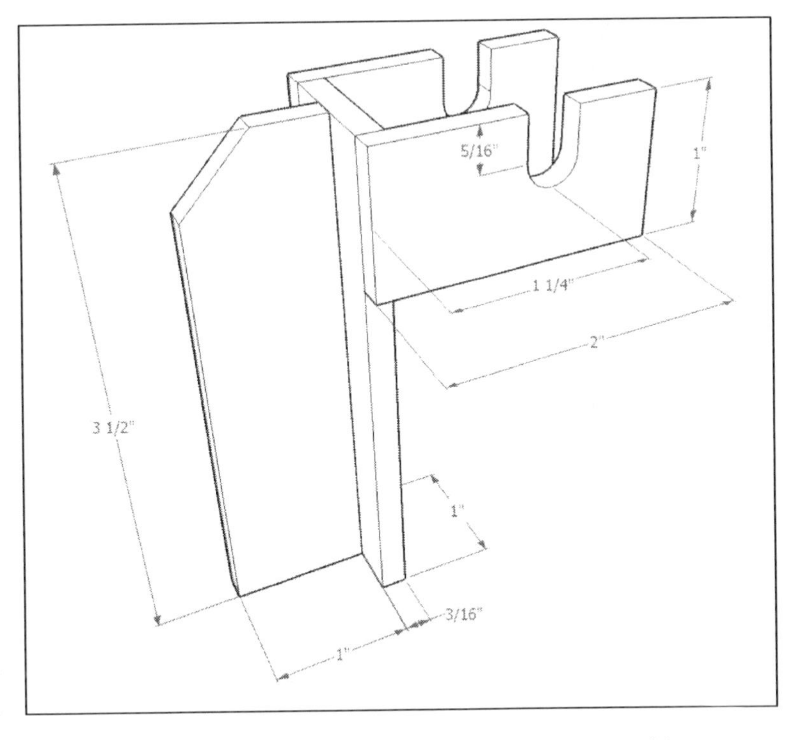


Figure 120 - The Crankshaft Bracket is made from acrylic sheet. This one was made from 3/16" material but yours can be made from the same 1/4" material used in the rest of the engine.

The slots in the top of the bracket are shaped to hold the glass beads that are used for bearings. The beads will be held in place with epoxy and must be aligned with each other so that the crankshaft can turn freely. I did not include that measurement on the drawing because the size of your beads may vary from mine.

Examine the measurements in the illustration and create some paper templates to assist you in cutting the acrylic parts. This bracket supports the glass bearings exactly 1" apart. Do not try to make the bracket any narrower. I have sometimes wished I made this bracket a little wider to simplify the assembly of the crankshaft. Placing the bearings about 1 1/2" apart will simplify things for you a little and will not interfere with the operation of the engine later.

The parts of the crankshaft bracket are joined with acrylic cement. The bearings can be attached to the bracket before or after the bracket is attached to the engine.

I used acrylic cement to attach the bracket to the top of the engine. This works very well as both parts are made from acrylic sheet.

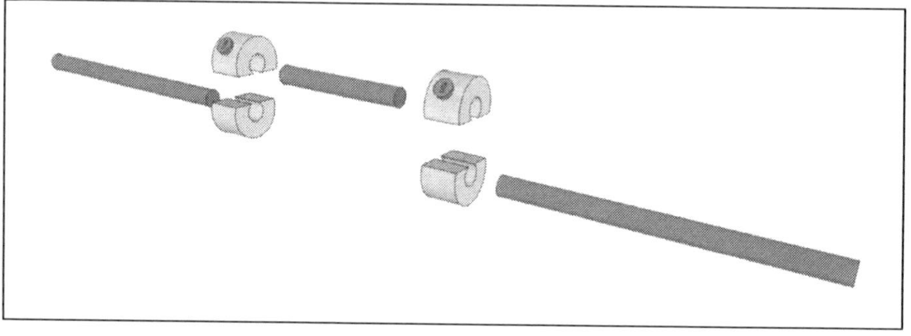

**Figure 121 - Crankshaft: Exploded View**

### *Make a Crankshaft*

Making the crankshaft was a difficult challenge for me. Several attempts were required before I achieved success. Hopefully you will be able to capitalize on my "research" and get it done on the first attempt. I designed the crankshaft with these requirements:

- It needs to be made from 1/8" music wire in order to use a glass bead as the pushrod bearing.
- It needs to have an offset of slightly less than 1/8".
- Both ends of the driveshaft must be true to each other (In other words, they must be aligned well).
- The center crank section needs to be parallel to the outer driveshaft sections.
- It must be something that can be assembled and disassembled on the engine without removing the bearings.

The solution is to use steel shaft collars to create the offset of the crankshaft. I used a belt sander and a simple jig to sand one side of the shaft collars flat, and then soldered them together in pairs. Here is the step by step process:

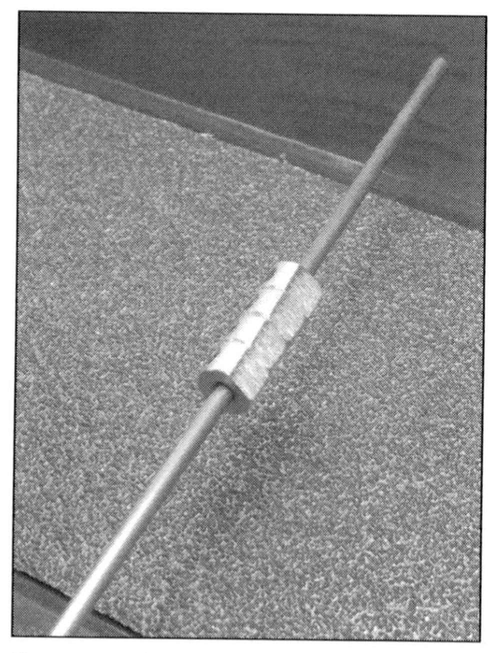

**Figure 122 - Check your work frequently to insure that you are removing equal amounts of material from all 4 shaft collars.**

- Attach four identical shaft collars to piece of 1/8" music wire that is at least 8" to 10" long. It needs to be long because it will get hot during the sanding/grinding process. This process will destroy the music wire.

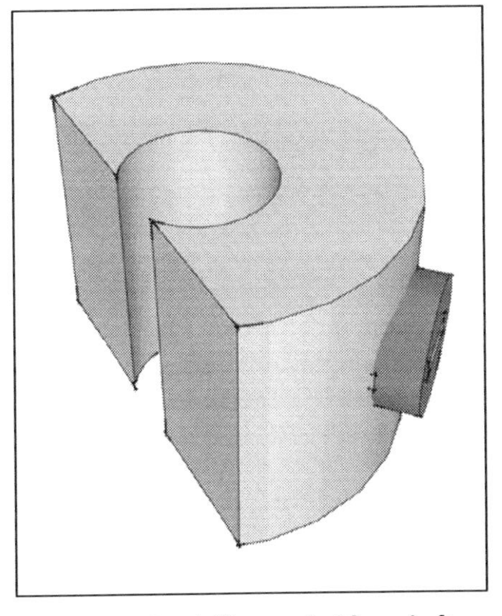

Figure 123 - Sand, file, or grind four shaft collars until the center hole is exposed. Try to make all four parts identical.

- Use a file, a belt sander, or a grinder to make an identical flat side on all four shaft collars at the same time. Align the flat side so that it is about 45° away from the set screws of the shaft collars.
- Grind or file until the music wire is just barely exposed. You want to grind off enough material so that you just start to expose the center hole of the shaft collar. This will provide you with the required crankshaft offset of just less than 1/8". Sanding or grinding to this point will also put a small flat surface on the side of the music wire. The wire will be discarded when you are finished making these parts.
- Check your work frequently to insure you are removing equal amounts of material from all four shaft collars.
- Do your best to make the four pieces identical. If you take them off of the music wire you can compare them to see if they are equal.
- Clamp two of the shaft collars together so that the flat faces you just made are face to face. Heat them carefully with a propane torch and apply a little solder to the joint. It only requires a small drop of solder. When the steel is hot enough the solder will flow into the joint. It is exactly the same process a plumber uses to sweat solder a pipe joint.
- Be careful you do not solder the set screws. If you do happen to get some solder on the set screws you can sometimes salvage them by using your Allen wrench to keep the screw in motion as the joint cools.

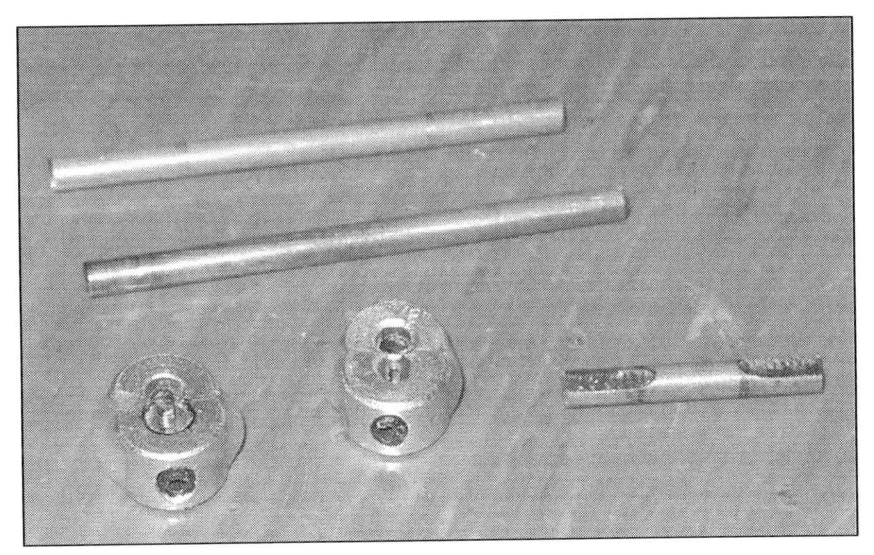

Figure 124 - Crankshaft parts are ready for assembly. The flat spots on the center section allow it to fit in the shaft collars.

The next step is to make the shaft sections for the crankshaft. The outboard pieces are 2" long and the inboard piece is 15/16" long. All are made from 1/8" music wire. The inboard piece needs to be slightly shorter than the distance between your bearings, so measure your

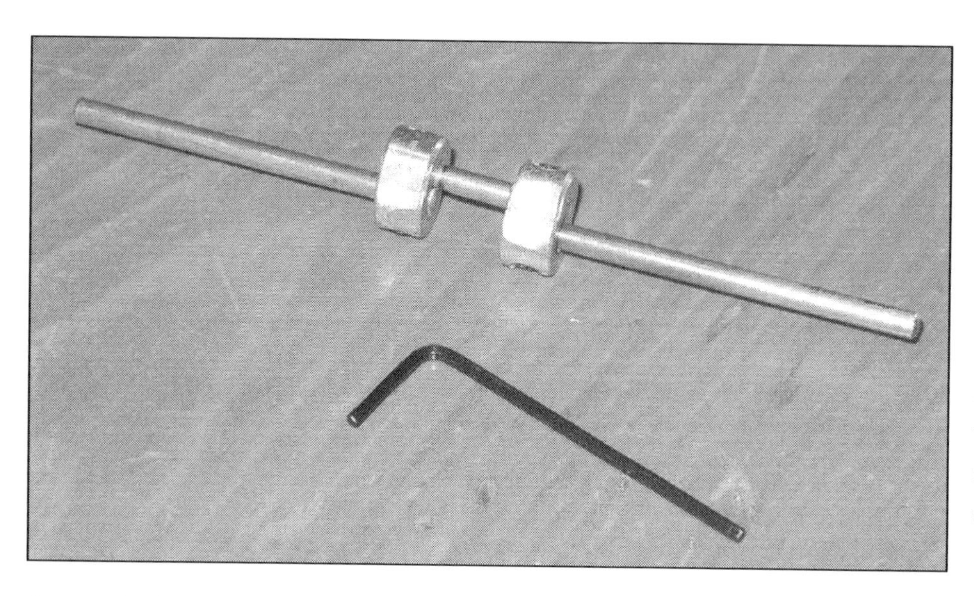

Figure 125 - The assembled crankshaft. The flat surfaces on the center section are touching the longer shaft of the outboard segment. This one appears to meet all our specifications.

bracket carefully and cut this piece to match the space available. (If your bearings are 1 1/2" apart, the center shaft should be about 1 7/16" long.) Polish the ends of the shaft after cutting with some fine sandpaper or a file so that it is easy to slip them into the shaft collars.

Because we ground the shaft collars down to the point where the center hole was exposed, you will not be able to fit two pieces of music wire into the holes on the joined shaft collars. You must remove some material from the side of one of the shafts to make room. The photos show how to grind some small flats on the side of the center shaft section. That allows you to insert two pieces into your new double shaft collar. This is also what creates the crank offset of just under 1/8"

Finally, assemble the crankshaft and check to see if it meets the requirements.

### *Build the Magnet Arms*

The engine has two rotating arms. Each arm carries a magnet. The magnet is set to oppose the magnets inside the pressure chamber on the displacer. As each magnet passes through the bottom of its arc it pushes the displacer to the opposite side of the pressure chamber.

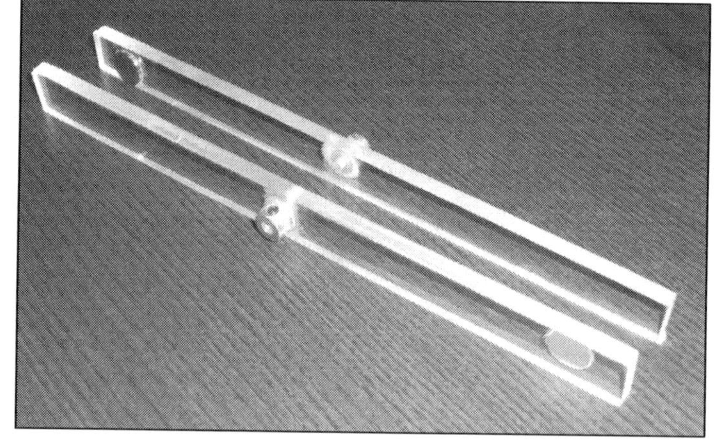

Figure 126 - The magnet arms each have one magnet and a shaft collar attached. The magnets must be set to push the displacer magnets away.

The arms also act as the flywheel for this design. They must have enough mass to function as a flywheel, but if they are too heavy they will overload the bearings and friction will prevent the engine from running.

The dimensions of the magnet arms are 8 1/2" long, 11/16" wide, and 1/4" thick. Cut two pieces of 1/4" acrylic to these dimensions. Sand the edges if necessary to make them smooth and even.

Use sandpaper to rough up a spot at the center of each piece. Use epoxy to attach a steel shaft collar at the exact center of each piece. After the glue has cured, carefully drill through the center of each shaft collar with a 1/8" drill bit, using the shaft collar as a drill guide.

One magnet is mounted on each arm. The magnet is mounted on the side that will be facing in towards the engine body. Dry fit the parts to determine the proper magnet location. The magnets need to be aligned to match the height of the internal magnets on the displacer. The magnets must be set so that they oppose (push away from) the magnets on the displacer. Once you have determined the proper polarity and location for the magnets you can glue them into place with epoxy.

### *Build and Attach the Pushrod*

The pushrod is made from a short length of wire. It can be made from music wire or a softer craft. One end of the pushrod will be bent into an eye and a glass bead will be glued into the eye. The other end will be attached to two nuts that will be threaded onto the attachment bolt that is in the center of the diaphragm.

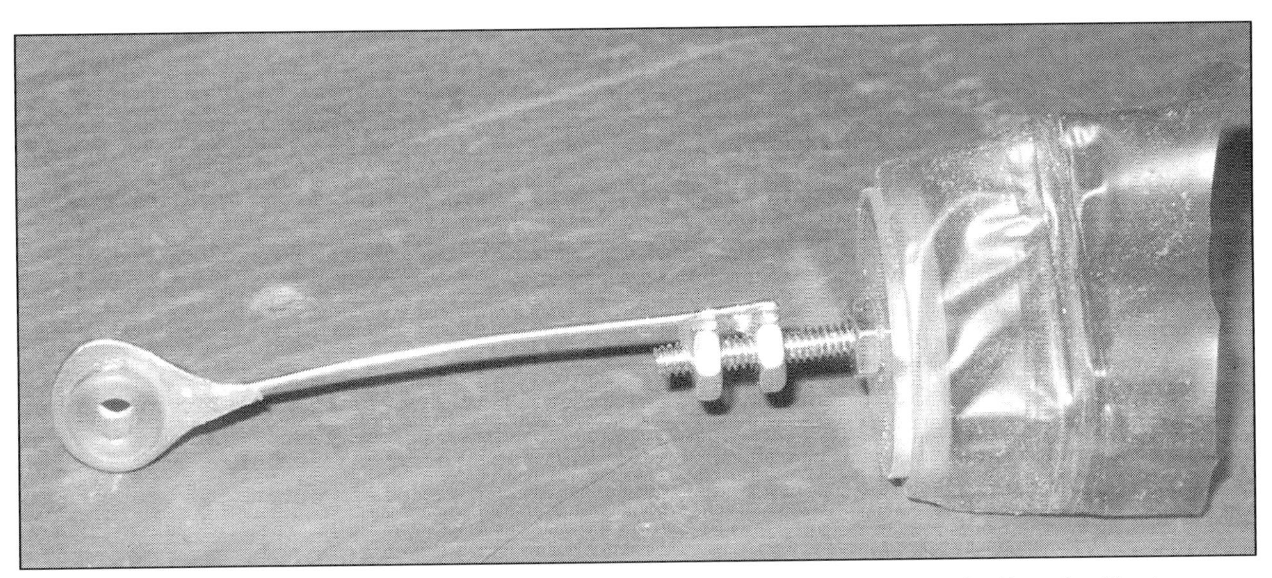

**Figure 127 - The pushrod has a glass bead attached at one end and two nuts glued to the other end. This pushrod has been attached to the diaphragm assembly and is ready to be mounted to the engine.**

**Note:** The other engines in the book used a slightly different method to connect the attachment bolt to the pushrod. That method will also work in this application.

The best way to determine the proper length of the pushrod is to assemble the diaphragm and the crankshaft and measure the distance between them. The piston needs to travel up and down inside the drive cylinder without being obstructed by the film can lid and without fully extending the diaphragm material. The pushrod length will be adjustable so there is a slight margin for error.

Make the loop and attach the glass bead bearing first. Use epoxy to glue the bead into place. After the glue has cured mount the pushrod onto the crankshaft, measure and trim it to length. After trimming you can remove it from the engine and attach two nuts to the other end of the wire. Thread two nuts onto the bolt and space them about 1/4" apart. Use epoxy to glue the pushrod to the flat sides of both nuts

### *Assemble the Drive Mechanism*

- Attach the diaphragm to the pushrod as illustrated above.
- Place the drive cylinder lid around the pushrod and set the diaphragm in place. Do not snap the lid down on the diaphragm just yet.
- Carefully assemble the crankshaft with the center section of the crankshaft going through the pushrod bearing. Adjust the shaft collars of the crankshaft so that the shaft is properly aligned.
- Gently hold the diaphragm in place and rotate the crankshaft to check the length of the pushrod. You can adjust the length of the pushrod (if required) by gently turning the diaphragm and screwing or unscrewing the joint between the diaphragm and the pushrod. The piston needs to travel freely inside the drive cylinder without hitting the lid at the top and without fully extending the diaphragm material at the bottom of its swing.

**Figure 128 - The assembled drive mechanism.**

- When the pushrod length is correct, snap the lid onto the top of the drive cylinder to seal the pressure chamber and hold the diaphragm in place.
- Install the magnet arms near the midpoint of each outboard axle. Check and make sure the magnets are set to repel the displacer as they pass by the pressure chamber in their rotation.

### *Set the Timing*

You must synchronize the engine by setting the timing for the crankshaft and magnet positions relative to the warm and cool sides of the engine.

- Determine which side of your engine will be the "warm side". I always set mine up so that the side without the stand braces is the warm side.
- Set the crankshaft so that the piston is elevated to the top of the stroke.
- Hold the crankshaft in the "up" position and rotate the magnet arm on the cold side of the motor so the magnet is down. It should be pushing the displacer away from itself, moving it to the warm side.

- Continue holding the crankshaft in the full "up" position and adjust the magnet on the warm side so that it is at the top. Tighten the set screws on both arms.
- Rotate the arms several times and observe the motion of the displacer. When the piston reaches the bottom of its stroke the displacer should move from the warm side to the cool side. When the piston reaches the top of its stroke the displacer should be moving from the cool side to the warm side.

### *Balance the Flywheel*

Open the vents so that there is very little resistance to flywheel rotation. Spin the magnet arms and observe how the weight of the crankshaft and piston cause their movement to be uneven. Add a little weight to one side of one of the magnet arms and adjust as necessary as a counterbalance for the crankshaft. My counterbalance is two 1" pieces of 1/8" music wire. I hold them on to the arm with masking tape so they are easy to adjust.

### *Pre-Flight Checklist*

Let's just double check to make sure everything is in order. Your engine should now be fully assembled and ready for the first running!

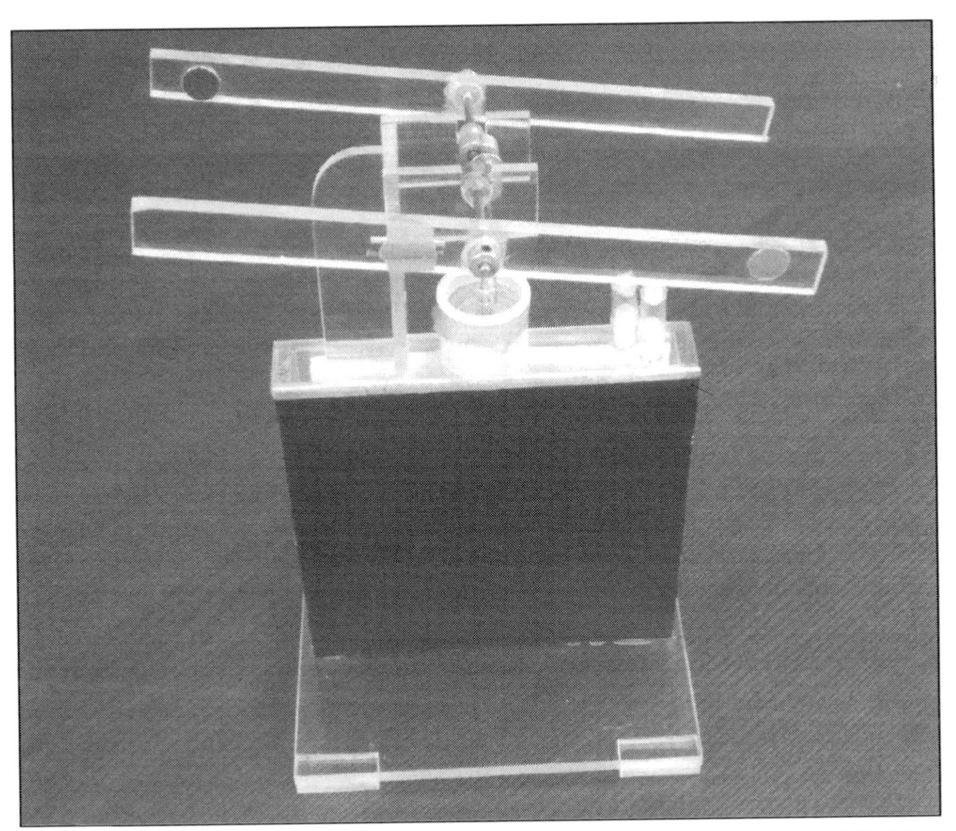

**Figure 129 - Engine #3 is complete and ready to run. You are looking at the "warm side" of the pressure chamber.**

You should now have an assembled Stirling engine with the following components:

- A sealed pressure chamber mounted vertically on a stand. The aluminum plating on the sides is painted black.
- The displacer inside the pressure chamber can rock easily from one side to the other. If you swing the magnetic swing arm close to the top of the pressure chamber it will push the displacer to the opposite side.
- There are two arms attached to the end of the crankshaft. Each arm has one magnet that is set to push the displacer to the opposite side of the pressure chamber.
- The timing has been set.
- The drive cylinder is attached to the top of the pressure chamber.
- The drive assembly (diaphragm, piston, and pushrod) are installed and adjusted correctly.
    - The piston can travel up and down inside the drive cylinder and is not obstructed by the film can lid or by the diaphragm material.
- Close the vents when the piston is in neutral position.

## Operation and Fine Tuning

Following the directions for setting the timing will put you in a good starting position for the first running of your engine. You will probably need to make some additional adjustments to get it to run smoothly. If all goes well it will start and run at between 60 and 100 RPM with a temperature differential of about 50°. Then, with care and fine tuning, you will be able to run it at smaller temperature differentials.

I sometimes lubricate the bearings with a small amount of silicone or WD-40®. Spray a small amount of lubricant onto a Q-tip and dab a small amount at the point where the axles pass through the bearings. Do the same for the bearing on the center of the crankshaft.

The easy way to get your engine started is to shine a 25 to 60 watt incandescent bulb a few inches from the warm side of the engine. Place an ice pack or a bag of frozen peas on the cold side of the engine. This will increase the temperature differential and help you get started with the tune-up process. After you have some time to observe and adjust your drive mechanism you will be able to run the engine without any ice.

The engine will warm up in 1 to 2 minutes and will need a slight push to get started. Rotate the magnet arms in either direction to start the engine. This engine will run in either direction. If it does not start running right away, allow another minute or two for the engine to warm up.

There are a variety of adjustments that may be necessary to get the engine to run continuously on its own. Here are some of the more obvious adjustments I often have to make.

- If the magnet fails to push the displacer to the other side it is either because the magnet is too far away from the displacer or it is traveling too fast. Try rotating the arm slower. If it still does not move the displacer loosen the set screw and slide the magnet arm closer to the engine body.
- If you can't get the magnet close enough to the engine body to move the displacer your magnet may be too weak. You can increase the magnet's strength by stacking another magnet on top of the magnet on the swing arm.
- If the magnet arm stops with the magnet at the bottom of the turn and is attracted to the displacer, your polarity is incorrect. You may have mounted the arms on the wrong side, or you may have mounted one of the magnets backwards.
- If the displacer is pushed away properly but the motor reverses direction, the magnet swing arm is too close to the engine body. Loosen the set screw and move the arm farther away from the engine body.

There is a very narrow margin of adjustment for successfully running this engine. The engine will run faster on higher temperature differentials. The swing arms must be closer to the engine body in this case or they will not move the displacer.

The arms need to be farther away from the engine body for running on lower temperature differentials.

I mentioned earlier I invested about $20 in a small infrared thermometer. I can point it at the surface of the engine and get an immediate temperature reading. This engine should idle along well for you the first time with a temperature differential of 50°. After you do some fine tuning it should consistently start and run with a temperature differential of 30° or less. When fine tuning is completed this engine will run with a temperature differential of 20° or so. The engine I built as a demonstration for this chapter will run down to a temperature differential of 20°. If the pressure chamber is filled with helium it will improve the performance even more.

I should point out that I have built several of these and I don't get the exact same performance every time. I have pointed out many times that there are variables in the building process that can have a big impact on your final results.

The trick to tuning a hand built engine is to observe it carefully and make only one adjustment at a time. If it improves performance, keep it. If it doesn't help, change it back. If it degrades performance, try changing it the other way. I always watch the movement of the diaphragm as I tune the engine. It is usually easy to tell what the engine is doing by watching the pressure changes on the diaphragm.

One of my favorite ways to run these engines is to set them in the sunshine. On a 70° day in the Pacific Northwest the sun will heat the warm side of the engine to as much as 120°, providing a 50° temperature differential. This works better than artificial heat sources, such as an incandescent light.

### Running from the Heat of Your Hand

I have been successful getting this engine to run from the heat of my hand only when the pressure chamber was filled with helium and the room temperature was at or below 68°. It took a lot of patience to get it to work in those conditions. I still consider this to be a low temperature differential engine, but I don't attempt to run it from the heat of my hand very often. It runs quite well at temperature differentials created by sunshine or a light bulb.

I still have hope that with some refinement this design could be improved upon and eventually be a true heat-of-the-hand engine. I have some new ideas I will try the next time I put one of these together. (I really want to try a shorter throw on the crankshaft and some better bearings...)

### Adding Helium

There are 2 vents in case you want to try running your engine with helium. I recommend you start using your engine with air. This will help you get a feel for how to tune it and make it work. Then add helium to see the performance difference it makes. To add helium, tip the engine so that the vents are on the bottom edge of the pressure chamber. Open both vents. Connect one vent to a helium balloon and discharge about 1/2 of its contents into the engine, then seal the pressure chamber by closing both vents. This process has now flushed most of the room air out of the engine and replaced it with the lighter helium.

When running on air or helium it is sometimes necessary to open one vent briefly so that the interior of the engine can equalize pressure with the environment. Changes in the weather bring changes in barometric pressure. Your pressure chamber is a crude barometer and is sensitive to these atmospheric pressure changes. If you are careful you can equalize the pressure several times before you notice a loss of performance caused by a loss of helium.

# Appendices

## A: Tips to Improve Engine Performance

Designing and building these engines has been a fun challenge and an interesting hobby. One of the most difficult challenges for me in writing this book was to be able to stop making changes and improvements to the designs long enough to document the construction of each model.

Here are a couple of recent modifications I have made to Engine #2 that seemed to make it run smoother and be easier to adjust.

The first change I made was to shorten the height of the flywheel. I measured carefully and trimmed about 1" off of the axle that holds the flywheel. The goal is to get the weight of the flywheel closer to the bearing. The flywheel is now just about 1/4" above the top of the drive cylinder.

The second change I made was to replace the straight upper pushrod. I crafted a new upper pushrod that is bent in a manner that ads some spring action. That made it much easier to tune the engine. It is now a bit more forgiving and will run at a wider variety of settings.

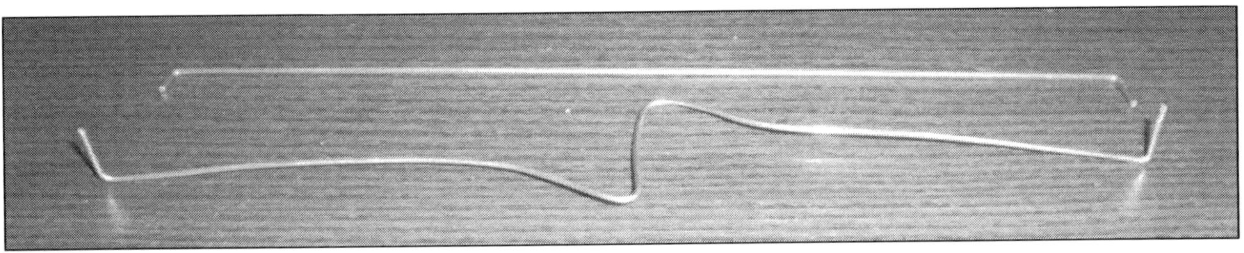

**Figure 130 - The original straight pushrod has been replaced with a new bent pushrod that has some spring to it.**

I trust that as you and other Stirling engine fanatics start building these designs that you will continue to build on this work and make these engines run better, faster, and stronger.

I will continue to use the Internet to keep you posted regarding changes and improvements as these designs evolve. Stay up to date on these and other Stirling engine projects by visiting me at: http://Stirlingbuilder.com.

## B: Proper Running Technique

Figure 131 - Here is how to run one of these engines with an electric lamp. This fixture is holding a 60 watt incandescent bulb a few inches from the warm side of the pressure chamber. The lamp shade is made of aluminum and does not interfere with the magnets.

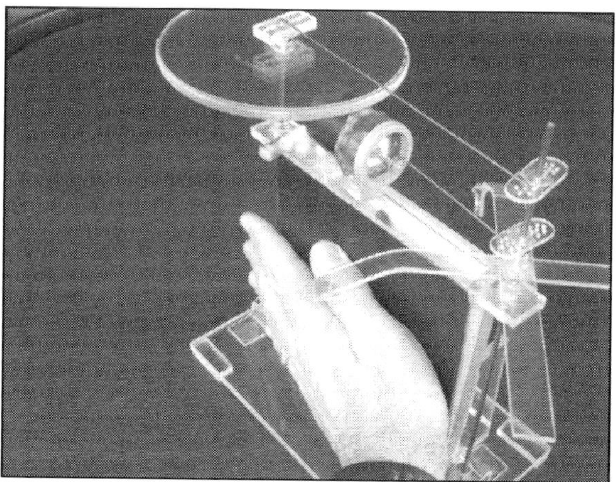

Figure 132 - Here is how to place your hand on the warm side of the engine. The engine will run after your hand warms up the side plate.

## C: Saw Blades and Techniques for Cutting Acrylic – Additional Information

Circular Saw Troubleshooting

Problem: Melting or Gummed Edges

Suggested Solutions:

1. Increase blade tooth size
2. Reduce saw speed
3. Increase feed rate
4. Use air to cool blade
5. Use a blade lubricant
6. Inspect blade for sharpness
7. Check blade/fence alignment
8. Reduce number of sheets in stack

Problem: Chipping

Suggested Solutions

1. Decrease blade tooth size
2. Provide better clamping and/or support for sheet stack
3. Reduce feed rate
4. Check blade and arbor for wobble
5. Inspect blade for sharpness

Band Saw Troubleshooting
Problem: Melting or Gummed Edges
Suggested Solutions:

1. Increase tooth size
2. Reduce saw speed
3. Use air to cool blade
4. Check blade sharpness

Problem: Chipping

Suggested Solutions:

1. Decrease tooth size
2. Slow down stock feed rate
3. Provide better clamping and/or support to eliminate vibration
4. Check blade for sharpness

(Trouble shooting tips courtesy of Sheffield Plastics Inc.)

## D: Drills and Drilling Techniques for Acrylic – Additional Information

- Standard wood bits can sometimes be used successfully in acrylic. Drill a pilot hole with a 1/16" bit and gradually step up drill bits until the desired hole size is reached. Step up in 1/64" increments if possible.
- Lubricate drill bits with light machine oil to help them run cooler.
- Running drill bits can be cooled with a stream of compressed air.

### *More Fine Print*

## E: Metric Conversions

| Fractional Inches | Decimal Inches | Metric mm |
|---|---|---|
| 1/32 | 0.0313 | 0.7938 |
| 1/16 | 0.0625 | 1.5875 |
| 3/32 | 0.0938 | 2.3813 |
| 1/8 | 0.1250 | 3.1750 |
| 5/32 | 0.1563 | 3.9688 |
| 3/16 | 0.1875 | 4.7625 |
| 7/32 | 0.2188 | 5.5563 |
| 1/4 | 0.2500 | 6.3500 |
| 9/32 | 0.2813 | 7.1438 |
| 5/16 | 0.3125 | 7.9375 |
| 11/32 | 0.3438 | 8.7313 |
| 3/8 | 0.3750 | 9.5250 |
| 13/32 | 0.4063 | 10.3188 |
| 7/16 | 0.4375 | 11.1125 |
| 15/32 | 0.4688 | 11.9063 |
| 1/2 | 0.5000 | 12.7000 |
| 17/32 | 0.5313 | 13.4938 |
| 9/16 | 0.5625 | 14.2875 |
| 19/32 | 0.5938 | 15.0813 |
| 5/8 | 0.6250 | 15.8750 |
| 21/32 | 0.6563 | 16.6688 |
| 11/16 | 0.6875 | 17.4625 |
| 23/32 | 0.7188 | 18.2563 |
| 3/4 | 0.7500 | 19.0500 |
| 25/32 | 0.7813 | 19.8438 |
| 13/16 | 0.8125 | 20.6375 |
| 27/32 | 0.8438 | 21.4313 |
| 7/8 | 0.8750 | 22.2250 |
| 29/32 | 0.9063 | 23.0188 |
| 15/16 | 0.9375 | 23.8125 |
| 31/32 | 0.9688 | 24.6063 |
| 1/1 | 1.0000 | 25.4000 |

| Temperature Differential Conversion | |
|---|---|
| Degrees Fahrenheit | Degrees Celsius |
| 1 | 0.56 |
| 2 | 1.11 |
| 3 | 1.67 |
| 4 | 2.22 |
| 5 | 2.78 |
| 6 | 3.33 |
| 7 | 3.89 |
| 8 | 4.44 |
| 9 | 5.00 |
| 10 | 5.56 |
| 11 | 6.11 |
| 12 | 6.67 |
| 13 | 7.22 |
| 14 | 7.78 |
| 15 | 8.33 |
| 16 | 8.89 |
| 17 | 9.44 |
| 18 | 10.00 |
| 19 | 10.56 |
| 20 | 11.11 |
| 21 | 11.67 |
| 22 | 12.22 |
| 23 | 12.78 |
| 24 | 13.33 |
| 25 | 13.89 |
| 26 | 14.44 |
| 27 | 15.00 |
| 28 | 15.56 |
| 29 | 16.11 |
| 30 | 16.67 |

7277826R0

Made in the USA
Lexington, KY
08 November 2010